Intelligent Systems Reference Library

Volume 274

The aim of this series is to publish a Reference Library, including novel advances and developments in all aspects of Intelligent Systems in an easily accessible and well structured form. The series includes reference works, handbooks, compendia, textbooks, well-structured monographs, dictionaries, and encyclopedias. It contains well integrated knowledge and current information in the field of Intelligent Systems. The series covers the theory, applications, and design methods of Intelligent Systems. Virtually all disciplines such as engineering, computer science, avionics, business, e-commerce, environment, healthcare, physics and life science are included. The list of topics spans all the areas of modern intelligent systems such as: Ambient intelligence, Computational intelligence, Social intelligence, Computational neuroscience, Artificial life, Virtual society, Cognitive systems, DNA and immunity-based systems, e-Learning and teaching, Human-centred computing and Machine ethics, Intelligent control, Intelligent data analysis, Knowledge-based paradigms, Knowledge management, Intelligent agents, Intelligent decision making, Intelligent network security, Interactive entertainment, Learning paradigms, Recommender systems, Robotics and Mechatronics including human-machine teaming, Self-organizing and adaptive systems, Soft computing including Neural systems, Fuzzy systems, Evolutionary computing and the Fusion of these paradigms, Perception and Vision, Web intelligence and Multimedia.

Indexed by SCOPUS, DBLP, zbMATH, SCImago.

All books published in the series are submitted for consideration in Web of Science.

Satchidananda Dehuri · Monalisa Jena ·
Sarat Chandra Nayak · Margarita N. Favorskaya ·
Smaranda Belciug
Editors

Advances in Quantum Inspired Artificial Intelligence

Techniques and Applications

Springer

Editors
Satchidananda Dehuri
Department of Computer Science
Fakir Mohan University
Balasore, Odisha, India

Sarat Chandra Nayak
Computer Science and Engineering
GITAM University
Hyderabad, Telangana, India

Smaranda Belciug
Department of Computer Science
University of Craiova
Craiova, Dolj, Romania

Monalisa Jena
Department of Computer Science
Fakir Mohan University
Balasore, Odisha, India

Yonsei University
Seoul, South Korea

Margarita N. Favorskaya
Institute of Informatics
and Telecommunication
Reshetnev Siberian State University
of Science and Technology
Krasnoyarsk, Russia

ISSN 1868-4394 ISSN 1868-4408 (electronic)
Intelligent Systems Reference Library
ISBN 978-3-031-89904-1 ISBN 978-3-031-89905-8 (eBook)
https://doi.org/10.1007/978-3-031-89905-8

This Springer imprint is published by the registered company Springer Nature Switzerland AG
The registered company address is: Gewerbestrasse 11, 6330 Cham, Switzerland

Preface

In recent years, the field of Artificial Intelligence (AI) has witnessed a remarkable surge in advancements, propelling the technology to unprecedented heights. One notable stride involves the refinement of machine learning algorithms, particularly in the realm of deep learning, enabling AI systems to discern complex patterns and make sophisticated decisions across diverse domains. One of the most interesting frontiers in AI has been the integration with quantum computing, marking a paradigm shift in the capabilities of intelligent systems. The synergy between AI and quantum computing has unlocked unprecedented computational power, promising to revolutionize the way we approach complex problem-solving. The ability of quantum computing to process information simultaneously across multiple states aligns seamlessly with the demands of AI algorithms, particularly those involved in optimization, machine learning, and pattern recognition. Researchers are exploring quantum-enhanced machine learning models that can handle vast datasets exponentially faster than classical counterparts. Moreover, the advent of quantum neural networks holds the potential to transform the very architecture of AI, presenting novel approaches to learning and decision-making. As these two transformative fields continue to intertwine, the integration of AI and quantum computing stands as a beacon of innovation, promising breakthroughs that could redefine the boundaries of what AI systems can achieve.

In this book, we have included 11 chapters starting from fundamentals of quantum computing to development of quantum inspired AI techniques and their applications in various real-world problems. In Chap. 1, Jena et al. presented the integration of quantum computing and bioinspired algorithms to address challenges in optimization and machine learning. By leveraging the unique principles of quantum mechanics, the capability of bioinspired techniques such as genetic algorithms and neural networks can be enhanced, leading to significant performance improvements, particularly in high-dimensional problems with complex constraints. Alongside, this chapter highlights how the advancements create new paths for solving real-world problems in artificial intelligence, robotics, and beyond, while also identifying future research opportunities in this evolving area.

The game-changing impact of quantum machine learning on predicting stock markets is significant. Quantum computing with its principles of superposition and entanglement have the ability to handle complex calculations and analyze large datasets at once shows promising results. Sahoo et al., in Chap. 2, have presented a quantum variational circuit which is a classifier and is designed to optimize parameters and find the best solution to a problem which in their case is to learn pattern and trends.

In Chap. 3, Kamilya and Paty explored the revolutionary potential of Quantum Dot Cellular Automata (QCA) as a groundbreaking technology in electronic circuitry, which could succeed traditional CMOS systems. QCA stands out for its significant advantages, offering greater efficiency, higher operational speed, and lower power consumption, making it a compelling candidate for next-generation electronic circuitry.

Pimpalshende et al., in Chap. 4, examined the fascinating intersection of quantum computing and natural language processing, which has generated increasing interest in the new discipline of Quantum Natural Language Processing (QNLP). This topic covers a broad range of NLP activities and uses the capabilities of quantum mechanics to handle important language processing issues. Further, Kayal et al., in Chap. 8, embarked on a rigorous exploration of QNLP, commencing with a meticulous examination of quantum probability and cognition. By elucidating the advantages afforded by these quantum concepts over classical cognitive-based SA theories, they lay the groundwork for understanding the transformative potential of QNLP.

In Chap. 5, Behera and Jena provided a comprehensive overview of climate science fundamentals, reviews major QML techniques applied in climate science, and explores the synergies between quantum computing and machine learning in advancing sustainable climate solutions. Similarly, Kedia et al., in Chap. 9, provided an overview of the newly developed field of Quantum Machine Learning (QML) and the diversification of its fields of uses. When integrated with the principles of quantum computing, and machine learning, QML provides enhanced computational power, which is likely to change the ways various sectors handle data and solve problems. This chapter also offers an extended discussion of the current applications of QML in finance, healthcare, cybersecurity, and artificial intelligence.

Kumari et al., in Chap. 6, explored the current landscape of quantum cloud computing, highlighting key developments and ongoing challenges. Authors examine major advancements, including novel quantum algorithms, hybrid computing models, and improvements in quantum cloud infrastructure. Additionally, critical challenges such as security concerns, resource allocation complexities, and the integration of quantum technologies into existing cloud architectures are addressed. A comparative analysis of Infrastructure as a Service (IaaS), Platform as a Service (PaaS), and Quantum as a Service (QaaS) is also presented, offering insights into their distinct roles and capabilities. The discussion further extends to major quantum cloud platforms, such as IBM Quantum Experience and Amazon Braket, and their contributions to innovation and accessibility. This chapter provides valuable insights for researchers and industry professionals, enabling future advancements in quantum cloud computing.

Quantum Computing (QC) along with Edge-Driven Intelligence (EDI) is becoming a powerful paradigm for decentralized decision-making and data processing driven by the explosive growth of Internet of Things (IoT) appliances and the massive growth of data generated at the network edge. However, there are also a lot of security, privacy, and trust-related issues that come with executing EDI. Some of the most pressing issues our world is facing could be resolved by quantum computing, include those related to materials science, energy, climate, agriculture, health, environment, etc. As the size of the system increases, classical computing becomes more difficult to solve some of these issues. Thus, considering the security, privacy, and trust in the context of QC and EDI, in Chap. 7, Panda et al. delved into the main problems and factors encountered in present era of technology.

In current era of technology, the adoration of Internet of Things (IoT) is rising day by day owing to extended internet connectivity as well as embedded technological proliferation. In this context, data processing and simulations could be greatly accelerated by quantum computing, which could have significant applications in the medical field. Priyadarshani, in Chap. 10, put efforts to throws light on the latest illustrations of various implementations of IoT while addressing various technical issues associated with healthcare sector.

Deep learning has demonstrated remarkable success in various domains, yet the computational demands of training large neural networks continue to pose challenges. The research by Chiranjevi et al. investigates the integration of quantum computing principles into neural network architectures, aiming to explore and exploit the potential quantum advantages in deep learning tasks. Their study focuses on Quantum Neural Networks (QNNs), where quantum bits (qubits) are leveraged to encode and process information in quantum superposition. Quantum entanglement and parallelism offer unique possibilities for enhancing the expressiveness and computational efficiency of deep learning models.

The book is crafted for a diverse readership, from seasoned researchers and practitioners in quantum computing and AI to students and enthusiasts eager to grasp the essence of this transformative convergence. At last, we extend a very deep sense of gratitude to the authors who have contributed their valuable work to fulfill the goal of the book.

Balasore, India — Satchidananda Dehuri
Seoul, South Korea — Monalisa Jena
Hyderabad, India — Sarat Chandra Nayak
Krasnoyarsk, Russia — Margarita N. Favorskaya
Craiova, Romania — Smaranda Belciug

Contents

Chapter 1
Quantum-Enhanced Bioinspired Algorithms: An Overview of Optimization and Learning

Monalisa Jena, Sarat Chandra Nayak, and Satchidananda Dehuri

Abstract Quantum computing presents a transformative approach to solving complex optimization problems. It has the ability to increase the computational power by solving complex real-life problems far efficiently than classical methods. This chapter explores the integration of quantum computing and bioinspired algorithms to address challenges in optimization and machine learning. By leveraging the unique principles of quantum mechanics, the capability of bioinspired techniques such as genetic algorithms and neural networks can be enhanced, leading to significant performance improvements, particularly in high-dimensional problems with complex constraints. By conducting thorough theoretical analysis and analyzing case studies, we showcase how quantum-enhanced algorithms can tackle complex, real-world problems across various fields. We examine the role of quantum genetic algorithms in optimizing solution diversity and convergence, alongside the application of quantum computing to neural networks, which can accelerate training and enhance pattern recognition. The chapter highlights how these advancements create new paths for solving real-world problems in artificial intelligence, robotics, and beyond, while also identifying future research opportunities in this evolving area.

Keywords Quantum computing · Neural networks · Genetic algorithm · Bioinspired algorithms · Artificial intelligence

M. Jena (✉) · S. Dehuri
Department of Computer Science, Fakir Mohan University, Balasore, Odisha, India
e-mail: bmonalisa.26@gmail.com

S. C. Nayak
Department of Computer Science and Engineering, GITAM, Hyderabad, Telangana, India

S. Dehuri et al. (eds.), *Advances in Quantum Inspired Artificial Intelligence*, Intelligent Systems Reference Library 274, https://doi.org/10.1007/978-3-031-89905-8_1

1.1 Introduction

Bioinspired algorithms are a class of optimization techniques designed by mimicking different aspects of nature such as natural selection, brain-like processing, collective behavior, and cooperation. Bioinspired techniques have successfully addressed complex challenges across diverse fields like artificial intelligence, data science, optimization, pattern recognition, and cybersecurity [1]. Quantum computing is a seamless integration of the quantum mechanics concepts and computer science algorithms. It has the ability to redefine computation by offering exponential speedups for certain problems. This makes quantum computing a transformative tool for advancing bioinspired algorithms, especially in addressing challenges involving high-dimensional search spaces and complex constraints [2]. Genetic algorithm is one of the notable bioinspired technique because of its flexibility and resilience across a wide range of optimization challenges. Drawing inspiration from the principles of natural selection and evolution, GA uses mechanisms such as selection, crossover, and mutation to refine solutions iteratively, adapting them over time to achieve optimal or near-optimal solutions. The recent integration of quantum computing with GAs has given rise to an advanced variation known as quantum genetic algorithm (QGA) [3]. By leveraging the distinctive features of quantum mechanics, QGAs significantly enhance the diversity within the population, accelerate the convergence process, and improve the overall quality of solutions.

Artificial neural network (ANN) is another fascinating bioinspired algorithm into which academicians are interested in incorporating quantum computing principles [4]. Quantum-enhanced neural networks can surpass the constraints of traditional ANNs, particularly in handling large datasets, and boosting the network's capacity to identify complex patterns. This chapter offers an exciting exploration of the powerful intersection between quantum computing and bioinspired algorithms, a topic that is becoming increasingly vital in today's rapidly evolving technological landscape. As quantum computing continues to evolve over time, its integration with bioinspired methods like genetic algorithms and neural networks holds the ability to transform the optimization and machine learning techniques. In this chapter, we aim to cover the fundamentals of quantum computing, bioinspired algorithms, and highlight the influence of quantum technologies on genetic algorithms and neural networks.

The paper is organized as follows: Sect. 1.2 introduces the foundational concepts of quantum computing, outlining its core principles and distinguishing features. Section 1.3 explores bioinspired algorithms, discussing their fundamental mechanisms and applications. Section 1.4 delves into the integration of quantum computing with genetic algorithms, examining how quantum principles enhance their efficiency and solution quality. Section 1.5 focuses on the application of quantum computing to neural networks, exploring its potential to accelerate training and improve pattern recognition. Finally, Sect. 1.6 presents the conclusion, summarizing key insights and identifying promising directions for future research in this evolving field.

1.2 Quantum Computing Basics

In the late 1990s, quantum computing captured the interest of scientists and researchers as they recognized the growing intersection of computer science and quantum mechanics. The fundamental distinction between classical and quantum computing lies in how information is stored and processed, as well as the underlying logic they employ. Unlike classical computers that rely on binary bits, quantum computers utilize qubits. Qubit is a basic unit of information in quantum computing. Qubits harness the principles of quantum mechanics, enabling them to exist in multiple states simultaneously, unlocking unprecedented computational potential. Qubits can be expressed as $|0\rangle$, $|1\rangle$, or as a superposition of both $|0\rangle$ and $|1\rangle$. The key quantum concepts are discussed as follows.

1.2.1 Superposition

The phenomenon that allows a qubit to exist in multiple states simultaneously is known as superposition. The most general state of a qubit is the superposition of its two basic states, $|0\rangle$ and $|1\rangle$ [5].

$$|\psi_1\rangle = x\,|0\rangle + y\,|1\rangle \tag{1.1}$$

Here, x and y are called amplitudes, that describe the proportion of qubit in $|0\rangle$ or $|1\rangle$. $x^2 + y^2 = 1$. In superposition, in addition to 0 and 1, the qubit considers all the states in between 0 and 1. In classical computers, when there are two bits, the possible combinations can be $\langle 00, 01, 10, 11\rangle$. However, in case of qubits, the superposition of the states can be in the following manner:

$$|\psi_1\rangle = x\,|00\rangle + y\,|01\rangle + z\,|10\rangle + w\,|11\rangle \tag{1.2}$$

1.2.2 Entanglement

It is the phenomenon of correlation of two or more qubits in such a manner that both are interdependent on each other. For two qubits $|00\rangle$ and $|11\rangle$, the possible entangled states can be [6]:

$$|\phi^+\rangle = \frac{1}{\sqrt{2}}(|00\rangle + |11\rangle) \tag{1.3}$$

$$|\phi^-\rangle = \frac{1}{\sqrt{2}}(|00\rangle - |11\rangle) \tag{1.4}$$

$$\left|\psi^{+}\right\rangle = \frac{1}{\sqrt{2}}(|01\rangle + |10\rangle) \tag{1.5}$$

$$\left|\psi^{-}\right\rangle = \frac{1}{\sqrt{2}}(|01\rangle - |10\rangle) \tag{1.6}$$

The entangled states are called Bell states [7]. The Bell state represents maximum entangled quantum states of two qubits. Here ϕ represents entangled state of quantum bits which share the same base state, such as $|00\rangle$ and $|11\rangle$, and ψ represents entangled state of quantum bits which share the different base state, such as $|01\rangle$ and $|10\rangle$. Here, the use of multiple entangled qubits can result in a significant computational speed-up in a quantum computer compared to the classical computers.

1.2.3 Interference

Quantum interference refers to a situation where the amplitudes of two qubits combine either constructively or destructively. In constructive interference, the amplitudes are in phase, causing their peaks and troughs to align. So, when the amplitudes are added with each other they result in a wave with a larger overall amplitude, and the probability of observing a particular outcome also increases. For example, if two quantum states have amplitudes $\psi_1 = x$ and $\psi_2 = x$, then their total amplitude will be $\psi = \psi_1 + \psi_2 = 2x$ and the probability will be: $P = |2x|^2 = 4|x|^2$.

In contrast to this, destructive interference occurs when amplitudes are out of phase, that means peak of one amplitude aligns with the trough of the other. As a result, the amplitudes cancel each other leading to a decreased or zero overall amplitude. For instance, if two quantum states have amplitudes $\psi_1 = x$ and $\psi_2 = -x$, then their total amplitude will be $\psi = \psi_1 + \psi_2 = x - x = 0$ and the probability will be: $P = \psi^2 = 0$.

1.2.4 Quantum Gates

Quantum gates differ from classical logic gates in that while classical gates such as AND, OR, and NOT operate on bits that are either 0 or 1, quantum gates act on qubits, which can exist in a superposition of both 0 and 1 simultaneously. Additionally, quantum gates can entangle qubits, enabling more intricate operations like quantum parallelism, which classical gates cannot perform. Quantum gates are represented by unitary matrices [8]. A matrix M is unitary *iff* $M^{-1} = M^{+}$, that means its inverse and conjugate transpose are equal. Some of the popularly used quantum gates are:

- **Hadamard Gate (H)**: It is mainly used to create superpositions. It is a single qubit state that transforms a qubit into an equal superposition of the $|0\rangle$ and $|1\rangle$ states. It is represented by the following matrix [9]:

$$H = \frac{1}{\sqrt{2}}\left(\begin{smallmatrix} 1 & 1 \\ 1 & -1 \end{smallmatrix}\right) \tag{1.7}$$

- **Pauli Gates**: These gates are a set of fundamental single-qubit quantum gates that form part of the Pauli group in quantum mechanics. Each of these gates applies a different transformation to the state of the qubit.
  - **Pauli-X gate**: It is a fundamental single-qubit quantum gate that performs a bit-flip operation. It flips the qubit state between $|0\rangle$ and $|1\rangle$. It can be represented in the form of the following matrix [10].

$$X = \begin{pmatrix} 0 & 1 \\ 1 & 0 \end{pmatrix} \tag{1.8}$$

  It means applying X to $|0\rangle$ results in $|1\rangle$, that is X $|0\rangle = |1\rangle$, and applying X to $|1\rangle$ results in $|0\rangle$, that is X $|1\rangle = |0\rangle$.
  - **Pauli-Y gate**: It performs a bit-flip operation combined with a phase shift of $\pi/2$ while flipping the qubit state. It can be represented in the form of the following matrix [10]:

$$Y = \begin{pmatrix} 0 & -i \\ i & 0 \end{pmatrix} \tag{1.9}$$

  It means when Pauli Y applied to a phae of i (a phase shift of $\pi/2$), it flips the qubit from $|0\rangle$ to $|1\rangle$. It can be viewed in the following equation:

$$Y|0\rangle = \begin{pmatrix} 0 & -i \\ i & 0 \end{pmatrix}\begin{pmatrix} 1 \\ 0 \end{pmatrix} = \begin{pmatrix} 0 \\ i \end{pmatrix} = i|1\rangle \tag{1.10}$$

  Similarly, when it is applied to a phae of $-i$, it flips the qubit from $|1\rangle$ to $|0\rangle$ along with the phase shift.

$$Y|1\rangle = \begin{pmatrix} 0 & -i \\ i & 0 \end{pmatrix}\begin{pmatrix} 0 \\ 1 \end{pmatrix} = \begin{pmatrix} -i \\ 0 \end{pmatrix} = -i|0\rangle \tag{1.11}$$

  - **Pauli-Z gate**: This gate applies a phase flip without changing the qubit's state. It applies a phase shift of π to state $|1\rangle$. That means, it multiplies the amplitude of $|1\rangle$ by -1. The Z matrix can be represented as follows:

$$Z = \begin{pmatrix} 1 & 0 \\ 0 & -1 \end{pmatrix} \tag{1.12}$$

$$Z|0\rangle = |0\rangle\,;\, Z|1\rangle = -|1\rangle\,. \tag{1.13}$$

1.3 Bioinspired Algorithms

Bioinspired algorithms are a set of computational methods designed by observing biological systems and natural processes. These are designed to solve real life complex optimization problems efficiently by using strategies used in nature. There are a plethora of bioinspired algorithms available in the nature [11], as nature serves as the treasure for such algorithms and techniques. They are broadly categorized into four main types:

- Evolutionary algorithms
- Swarm intelligence based bioinspired algorithms
- Multi-objective bioinspired algorithms
- Ecology based bioinspired algorithms.

Figure 1.1 shows a classification of different bioinspired algorithms, covering many techniques inspired by nature. In this section, we have tried to discuss some of the popularly used bioinspired algorithms from each category.

1.3.1 Genetic Algorithm

Genetic algorithms is one of the popularly used bioinspired algorithms designed on the principles of genetics and natural selection [12]. It belongings to the category of evolutionary algorithms. They are mainly used in optimization and searching techniques. The algorithm operates using a population of potential solutions. Each solution is called an individual and a set of attributes are called genes [13]. Through several operations, GAs evolve over time producing fittest individual for the next generation. Genetic algorithms are particularly useful for problems where the search space is vast and complex, such as optimization in engineering, scheduling, and machine learning. The key operations used in GA are:

- **Selection**: The process involves choosing the fittest individual for the subsequent generation by using a fitness function. This fitness acts as a criterion for evaluating the fitness of an individual. An individual with a higher fitness value is more likely to be selected. Commonly used methods for selection include roulette wheel selection, tournament selection, and rank-based selection.
- **Crossover**: It is also called as recombination. The pair of individuals selected in selection process are combined to produce offspring. Crossover can be single point, multi-point or uniform.
- **Mutation**: Mutation randomly alters an individual's genetic code, preserving genetic diversity within the population. Typical mutation methods include bit-flip mutation and random value mutation. Bit-flip mutation is applied to binary representations, while random value mutation is used for real-valued genes.

Fig. 1.1 Classification of bio-inspired algorithms

1.3.2 Artificial Neural Networks

ANNs are designed to process information through a network of interconnected units known as neurons. They have several unique features that make them powerful tools for a wide range of tasks. They exhibit non-linearity, allowing them to model complex relationships by using activation functions at each neuron. ANNs are highly adaptable, learning from data and improving their predictions over time. Their ability to process multiple inputs simultaneously through parallel processing enhances their efficiency, especially in deep learning models. In ANNs, knowledge is represented in a distributed manner across neurons and layers, providing robustness to noise and incomplete data. They also exhibit fault tolerance, as they can still function effectively even if some neurons or weights are damaged. Unlike conventional algorithms that follow fixed, predefined rules, ANNs excel at learning from data. Through iterative learning techniques like gradient descent, ANNs are capable of self-improvement, continually refining their performance. These characteristics make ANNs highly effective for applications in image recognition, speech processing, natural language understanding, and complex decision-making. Further insights about ANN will be explored in Sect. 1.5.

1.3.3 Ant Colony Optimization (ACO)

It is a widely used swarm-based bioinspired algorithm, that mimics the foraging behavior of ants to solve complex computational problems [14]. Just like ants use pheromones to find the shortest path between their living place and food, artificial ants in ACO explore possible solutions to a problem and find better solutions to a problem [15]. They deposit pheromones on better solutions following which other ants reach towards the solution. Over the time the collective intelligence of the ants converges to an optimal or near-optimal solution. The ACO algorithm is widely used in several applications like feature selection, clustering, job scheduling, routing and network optimization.

1.3.4 Multi-objective Artificial Bee Colony Algorithm

The artificial bee colony (ABC) algorithm is an optimization technique that simulates how bees search for food, share information, and communicate to find the best nectar sources. It solves optimization problems by imitating how bees locate food sources and share information within the hive. In ABC, each potential solution is treated as a food source, with its quality (fitness) evaluated based on the objective function being optimized. There are three kinds of bees: employed bees, responsible for finding a solution and exploring the surrounding area to find better solutions, onlooker bees who evaluate the solutions and probabilistically choose the best solution, and the third one, scout bees who randomly search for solutions when others cannot find a solution.

The multi-objective ABC algorithm extends the ABC algorithm to handle multi-objective optimization tasks by optimizing multiple conflicting objectives simultaneously. We present here a brief mathematical formulation of the multi-objective ABC algorithm. Suppose there are N objective functions to be optimized concurrently. Each solution $\mathbf{S} = (s_1, s_2, ..., s_n)$ is represented by a vector of decision variables in an n-dimensional space. The objective functions are expressed as: $g_1(\mathbf{S}), g_2(\mathbf{S}), ..., g_N(\mathbf{S})$, where $g_i(\mathbf{S})$ denotes the i-th objective function.

For each solution $\mathbf{S}_k$, the fitness is evaluated considering all the objectives. The fitness score $G(\mathbf{S}_k)$ is determined using Pareto dominance as follows [16]:

$$S_k' = S_k + \gamma(S_k - S_l) \tag{1.14}$$

Where $\mathbf{S}_k$ is the current solution, $\mathbf{S}_l$ is a randomly selected solution, and γ is a random step factor. During the onlooker phase, the probability $\mathcal{P}_k$ of selecting a solution is based on its fitness [17]:

$$\mathcal{P}_k = \frac{G(S_k)}{\sum_{i=1}^{M} G(S_i)} \tag{1.15}$$

Where, $G(\mathbf{S}_k)$ is the fitness of solution $\mathbf{S}_k$, and M is the total number of solutions. After a series of iterations, the algorithm forms a Pareto front $\mathcal{P}$ that includes all the non-dominated solutions, defined as: $\mathcal{P} = S_k | \nexists S_j$ such that S_j dominates S_k. The algorithm stops when no further improvements are detected in the Pareto front or after a predetermined number of iterations.

1.3.5 Biogeography Based Optimization (BBO)

This was introduced by Dan Simon in 2008 [18]. It draws inspiration from the way species are distributed in different environments, with each environment representing a possible solution to an optimization problem. The quality of a solution is assessed using a metric known as the habitat suitability index, where a higher index indicates a superior solution. In contrast to other evolutionary algorithms, BBO utilizes migration models that are based on biogeographic principles, enhancing exploitation and exploration of the search space. BBO has shown effective results in various fields, including engineering design, optimization of power systems, and machine learning, owing to its straightforwardness and capability to efficiently explore and exploit the solution landscape [19].

Quantum computing has the potential to enhance the performance of bioinspired algorithms by processing multiple solutions at the same time, accelerating the optimization process. For example, quantum genetic algorithms could use quantum superposition to represent multiple solutions in parallel, potentially improving search efficiency. Similarly, quantum versions of other bioinspired algorithms like PSO and ACO could enhance their ability to explore vast search spaces more effectively, accelerating convergence to optimal solutions. Quantum algorithms can address some of the challenges bioinspired algorithms face, such as computational cost and slow convergence, by exploiting quantum parallelism and interference.

Quantum computing has the ability to enhance a wide range of bioinspired algorithms, such as ACO and PSO. However, this chapter specifically focuses on the quantum genetic algorithm and quantum neural networks (QNN). These two algorithms are well-suited for integration with quantum computing due to their inherent capabilities to benefit from quantum principles like superposition and entanglement. By concentrating on QGA and QNN, we aim to highlight how quantum computing can optimize and accelerate the performance of these algorithms, offering unique advantages such as faster search spaces and more efficient learning processes. This chapter will delve into the specific applications and strengths of these two algorithms, leaving other bioinspired algorithms for future exploration.

In the following sections, we will discuss quantum genetic algorithm and quantum neural network in detail, examining their underlying principles, key algorithms, and potential real-world applications.

1.4 Quantum Computing for Genetic Algorithms

Quantum computing combined with genetic algorithm, popularly called QGA can solve the optimization and search problems much more efficiently. Traditional genetic algorithms represent solutions with binary strings, whereas QGAs utilize quantum bits that can be in superpositions of states, enabling a more effective search and exploration of the solution space. In this section, we are going to explore how genetic algorithm operations can be adapted in the context of quantum computing.

1.4.1 Quantum Crossover and Mutation

- **Quantum crossover**: Quantum Crossover involves the exchange of quantum states between two parent solutions to create offspring. In classical GAs, this would involve swapping parts of the binary strings of the parents. Unlike classical GA, QGA uses quantum gates for crossover operations. For instance, if we apply the CNOT gate to two qubits $|q_1\rangle$ and $|q_2\rangle$, the result will be [20]:

$$CNOT(|q_1\rangle \otimes |q_2\rangle) = |q_1\rangle \otimes (X^{q_1}\,|q_2\rangle) \tag{1.16}$$

where, X is the Pauli-X operator.
- **Quantum mutation**: It is used to change the states of qubits. It flips the qubit between $|0\rangle$ and $|1\rangle$, thereby introducing new solutions into the population. One most commonly used examples of mutation is Pauli X gate, explained in Sect. 1.2.4.

1.4.2 Quantum Measurement and Solution Selection

After quantum crossover and mutation takes place, a quantum measurement is performed to collapse the qubit superposition into a definite classical state, 0 or 1. The probability of measuring the state $|0\rangle$ or $|1\rangle$ is directly related to the amplitudes of the superposition. Mathematically, these probabilities are given by:

$$P(0) = |\alpha|^2\,, P(1) = |\beta|^2 \tag{1.17}$$

When the superposition collapses, a classical solution is obtained and evaluated based on its fitness value. The fitness is evaluated using fitness function $f(x)$. The best-performing solutions, based on their fitness, are then selected for the next generation.

1.4.3 Quantum Genetic Algorithm

A quantum GA operates by initializing a population of quantum solutions, represented as qubit chromosomes. These quantum solutions are then measured to generate candidate solutions evaluated based on their fitness, similar to traditional genetic algorithms. The best solution among the candidates is selected. Following this, genetic operations like crossover and mutation are applied on the chosen solutions to generate new offspring. The crossover process merges elements from two parent solutions to produce a new solution, whereas mutation adds random alterations to preserve diversity. Then quantum gates are applied to the qubits, updating their states and introducing further quantum-based variation into the population. Once the quantum states have been updated, the algorithm checks whether the termination condition is satisfied. This condition determines if the solution is good enough after a certain number of iterations. If the criteria are satisfied, the algorithm delivers the most optimal solution discovered so far and concludes. If the criteria are not satisfied, the process repeats, starting with evaluating the newly generated solutions. This cycle continues, with solutions evolving through quantum and genetic operations, until the stopping criteria are satisfied. The entire procedure is illustrated in Fig. 1.2.

1.4.4 Quantum Genetic Algorithm for the Traveling Salesman Problem: A Case Study

In this section, we explore a case study to dive deeper into the QGA concept, using the classic optimization problem, the traveling salesman problem (TSP), to demonstrate how it is solved with QGA.

Problem Definition:
Given a set of n cities, and the distances between each pair of cities, the objective is to find the shortest possible path that visits each city exactly once and returns to the starting city. The TSP can be formulated as [21]:

$$\min_{\sigma} \sum_{i=1}^{n-1} \delta(C(i), C(i+1)) + \delta(C(n), C(1)) \tag{1.18}$$

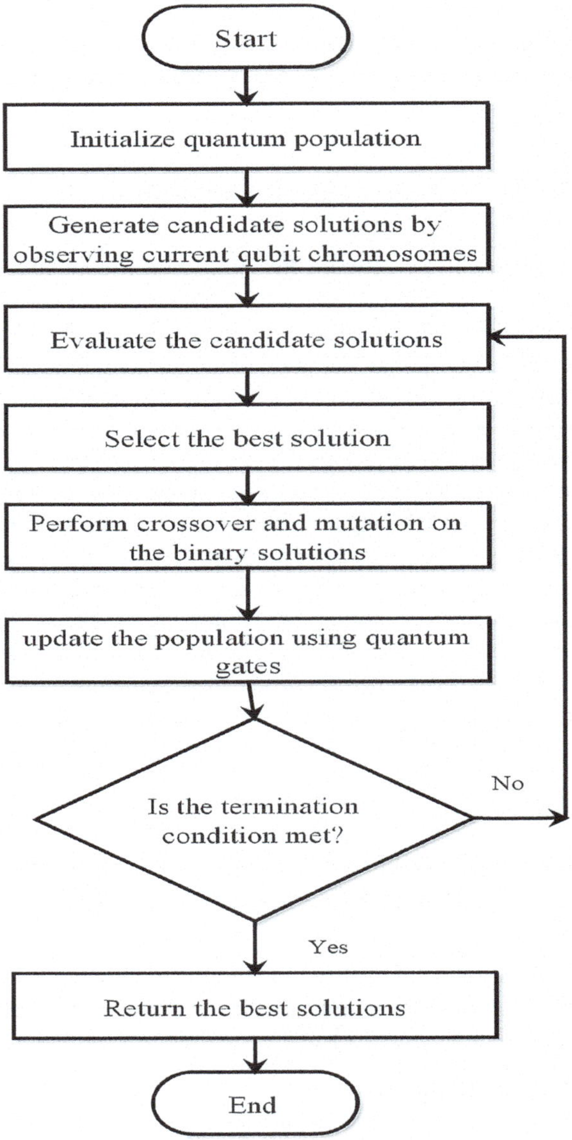

Fig. 1.2 Iterative process of quantum genetic algorithm

where, σ is a permutation of cities, $\delta(a, b)$ is the distance between cities a and b, $C(i)$ represents the i-th city in a given route, and the summation calculates the total distance traveled along the route defined by the permutation σ.

- **Quantum crossover for TSP**: In the QGA for TSP, the route or the solution can be represented by a quantum chromosome. Each individual city in the route will be represented by a qubit or a set of qubits in superposition. For n cities, the quantum state of the route can be represented as a superposition of all possible permutations, totaling $n!$ [22]:

$$|C\rangle = \sum_{i=1}^{n!} \alpha_i |P_i\rangle \tag{1.19}$$

where, $|P_i\rangle$ represents a specific permutation of cities, and α_i are the complex amplitudes for each permutation. For two possible paths $P_1 = (C_1, C_2, C_3, C_4)$ and $P_2 = (C_4, C_3, C_1, C_2)$, the *quantum crossover* for TSP combining parts of P_1 and P_2 can be represented as [23]:

$$|P_{offspring}\rangle = \frac{1}{\sqrt{2}} |(C_1, C_3, C_2, C_4) + (C_4, C_2, C_1, C_3)\rangle \tag{1.20}$$

Here the offspring is a superposition of the combined routes.

- **Quantum mutation for TSP**: For TSP, a mutation operation might swap the positions of two cities in the permutation. For example: One simple quantum mutation can be applied by using a quantum NOT gate (X-gate) to one of the cities in the permutation, causing a probabilistic change to that city's position. Let us consider the route C_1, C_2, C_3, C_4. Suppose mutation is applied to city C_2, and it is mutated to city C_3 probabilistically. This mutation transforms the quantum state by mixing the positions of C_2 and C_3 with a certain probability. After applying quantum mutation, the new quantum state $P_{mutated}$ may look like [24]:

$$|P_{mutated}\rangle = \frac{1}{\sqrt{24}} |(C_1, C_3, C_2, C_4) + (C_4, C_2, C_1, C_3) + (C_1, C_2, C_3, C_4) + \cdots\rangle \tag{1.21}$$

Here $\sqrt{24}$ represents the normalization factor. For 4 cities, we have $4! = 24$ possible routes.

- **Quantum measurement for TSP**: After applying the quantum mutation, the quantum state is measured, collapsing the superposition into one of the possible routes [25]. After measurement, the mutated route generated is (C_1, C_3, C_2, C_4).
- **Fitness function**: To evaluate the fitness of a solution to TSP, a fitness function is used. After applying the quantum crossover and mutation operations, the fitness is computed by collapsing the quantum state and measuring the corresponding distance for the specific permutation [26]:

$$F(\sigma) = \sum_{i=1}^{n-1} \delta(C(i), C(i+1)) + \delta(C(n), C(1)) \tag{1.22}$$

Using these fitness values, the algorithm identifies the best-performing individuals to create a mating pool for further refinement. The iterative process of evaluation terminates when a predefined stopping criterion is satisfied.

1.5 Quantum Computing for Neural Networks

Quantum computing can also be integrated into neural networks, called quantum neural network enhancing their capabilities through quantum superposition and entanglement. This combination enables neural networks to handle intricate, high-dimensional data more effectively, speed up optimization processes, and attain improved outcomes in pattern recognition and decision-making [27]. Traditional neural networks work well but often need a lot of resources to handle large tasks. QNNs leverage unique quantum properties to process information in innovative ways. In a QNN, qubits replace classical neurons, allowing for the representation of intricate data in higher-dimensional spaces via quantum states [28]. This capability introduces exponential parallelism, allowing QNNs to solve certain problems more efficiently than classical NNs. Combining quantum mechanics with neural networks helps create algorithms that work more like natural intelligence, making them particularly relevant for bioinspired computing.

This section delves into how quantum computing can be incorporated into neural networks, placing emphasis on the ways quantum principles can improve neural network models. By exploring the role of quantum computing in refining neural network architectures, we seek to demonstrate the potential of QNNs in tackling highly complex and resource-intensive problems. Integrating quantum computing with neural networks offers enhanced performance, greater efficiency, and increased capability, particularly for tasks involving large and complex data.

1.5.1 Key Quantum Concepts and Techniques Used in QNN

The key concepts and techniques that define QNN are discussed here. A QNN consists of quantum gates that transform qubits, much like classical neurons are activated in traditional neural networks. It uses gates like Pauli gates, Hadamard gate, discussed in Sect. 1.2.4. In addition to this, QNN involves several other techniques discussed as follows:

- **Quantum state encoding**: It converts classical data to quantum states, a necessary step for QNNs to process data. One prevalent encoding strategy is amplitude encoding. In this approach, a classical data vector, say $D = [d_1, d_2, d_3, .., d_m]$ is mapped to the amplitudes of a quantum state. The quantum state is represented as:

$$|\varphi\rangle = \sum_{j=1}^{m} d_j |j\rangle \tag{1.23}$$

For example, for a two-dimensional dataset $D = [0.2, 0.3]$, the quantum state will be:

$$|\varphi\rangle = 0.2|0\rangle + 0.3|1\rangle \tag{1.24}$$

- **Variational quantum circuits**: These are used to train QNNs by minimizing a loss function through a hybrid quantum-classical approach. A variational quantum circuit consists of parameterized quantum gates that are adjusted during the optimization process. The general form of a quantum circuit is:

$$U(\lambda) = \prod_{i=1}^{m} e^{-i\lambda_i H_i} \tag{1.25}$$

In this context, $U(\lambda)$ denotes the unitary matrix that models the quantum circuit, while λ_i represents the parameters that need to be optimized, and H_i are the Hamiltonians associated with various gates. Once the quantum circuit has handled the input data, an attempt is made to reduce the cost function $C(\lambda)$.

$$C(\lambda) = \langle\varphi(\lambda)|\, H\, |\varphi(\lambda)\rangle \tag{1.26}$$

- **Quantum backpropagation and optimization**: Quantum backpropagation updates the quantum circuit's parameters by computing the gradient of the cost function with respect to those parameters. While classical neural networks use backpropagation to minimize errors by adjusting weights, quantum neural networks employ quantum backpropagation to fine-tune the parameters of quantum gates through quantum gradients. For instance, let us consider the quantum parameterized state $|\varphi(\lambda)\rangle = cos(\lambda)\,|0\rangle + sin(\lambda)\,|1\rangle$, and the cost function as $C(\lambda) = \langle\varphi(\lambda)|\, H\, |\psi(\lambda)\rangle$, the gradient descent of the cost function with respect to λ will be:

$$\frac{\partial C(\lambda)}{\partial \lambda} = \frac{\partial}{\partial \lambda}(cos^2(\lambda)H_{00} + sin^2(\lambda)H_{11}) \tag{1.27}$$

1.5.2 *Different QNN Models*

In this section several QNN models proposed in the literature are discussed [29].

1.5.2.1 Quantum M-P Neural Network

This model was proposed by Zhou and Ding [30] after making a thorough analysis of the conventional McCulloch-Pitt (M-P) neural network. The M-P model, introduced in the 1940s, is one of the earliest models of artificial neurons [31]. It represents neurons as simple binary threshold units, where the output is 1 if the input exceeds a threshold and 0 otherwise. In the quantum M-P model, classical binary units are substituted with qubits. This enables quantum M-P networks to handle information with greater efficiency than conventional neural networks in specific applications.

Given inputs $x_1, x_2, x_3, ..., x_n$, and their associated weights $w_1, w_2, w_3, ...w_n$, the output to the M-P neural network is:

$$y = \begin{cases} 1, if \sum_{i=1}^{n} w_i x_i \geq \theta \\ 0, if \sum_{i=1}^{n} w_i x_i < \theta \end{cases} \tag{1.28}$$

where, θ is the threshold value. In case of quantum M-P model, the output is based on the quantum state of the qubit. For n input qubits $|x_1\rangle, |x_2\rangle, |x_3\rangle, ..., |x_n\rangle$, the total state of the neuron is the superposition of weighted inputs. The weighted sum S of the inputs is computed as:

$$S = \sum_{i=1}^{n} w_i |x_i\rangle \tag{1.29}$$

The output of the quantum M-P neuron depends on a quantum measurement, where the measurement outcome is determined by the probability amplitude of the quantum state. For two qubits $|x_1\rangle$ and $|x_2\rangle$, the input can be in one of the states $|00\rangle$, $|01\rangle$, $|10\rangle$ and $|11\rangle$. The output Y can be represented as:

$$Y = w_1\phi_1(x_1, x_2) + w_2\phi_2(x_1, x_2) + w_3\phi_3(x_1, x_2) + w_4\phi_4(x_1, x_2) \tag{1.30}$$

where, $\phi_1, \phi_2, \phi_3, \phi_4$ are basis functions used for mapping the quantum states to specific operations or behaviors, and w_1, w_2, w_3, w_4 are the corresponding weights associated with the basis functions.

If the quantum states are orthogonal, then the output of the model will be:

$$Y_k = \sum_{j=1}^{2^n} w_{kj} |a_1, a_2, ..., a_n\rangle \tag{1.31}$$

Where, Y_k is is the output for the k-th element of the quantum M-P model, w_{kj} are the weights associated with the different quantum states, and $|a_1, a_2, ..., a_n\rangle$ are the quantum basis states, which are orthogonal to each other. These states represent all possible combinations of n qubits, where $a_1, a_2, ..., a_n$ can be 0 or 1. The states $|00\rangle$ and $|01\rangle$ are orthogonal to each other as their scalar product is zero. For instance, for $n = 2$, Y_k in equation (30) becomes:

$$Y_k = w_{k1} |00\rangle + w_{k2} |01\rangle + w_{k3} |10\rangle + w_{k4} |11\rangle \tag{1.32}$$

Where, $w_{k1}, w_{k2}, w_{k3}, w_{k4}$ are weights associate with the states $|00\rangle$, $|01\rangle$, $|10\rangle$, $|11\rangle$, respectively.

For models where states are not orthogonal, that means the states are not mutually distinct and their inner product is not zero, for example: $\langle a_0|b_1\rangle \neq 0$, the output will be:

$$Y_{km} = \sum_{j=1}^{2^n} w_{kj} \left\langle j|m|\right\rangle \tag{1.33}$$

Where, Y_{km} is the output of the quantum model for the k-th and m-th states, w_{kj} is a weight factor associated with the quantum state, $\langle j\rangle$ and $\langle m\rangle$ represent two quantum states; $\langle j\rangle$ corresponds to the set of all possible quantum states and $\langle m\rangle$ corresponds to the output state, $\langle j|m|\rangle$ is the scalar product between the two states $\langle j\rangle$ and $\langle m\rangle$.

1.5.2.2 Quantum Inspired Neural Network (QINN)

This model was proposed by Menneer and Narayanan in 1995 [32]. There are two types of QINN: normalization QINN, which uses only quantum neurons and hybrid QINN which which combines classical and quantum neurons. In the hybrid QINN architecture, the input layer consists of n classical neurons, the hidden layer contains p quantum neurons, and the output layer consists of m classical neurons. The relationship between inputs and outputs is described by the following equations [33]:

$$h_j = B_j \sum_{i=1}^{n} x_i R_{ij} \left|\phi_{ij}\right\rangle \tag{1.34}$$

$$y_k = g\left(\sum_{j=1}^{p} w_{jk} h_j\right) \tag{1.35}$$

Where, x_i is the ith input value, R_{ij} is a quantum rotation gate applied to $\left|\phi_{ij}\right\rangle$, w_{jk} represents the weight connecting the jth hidden neuron to the k-th output neuron, $g(.)$ is an activation function. The error function for the output layer is given as:

$$E = \frac{1}{2}\sum_{k=1}^{m} \left(\hat{o_k} - o_k\right)^2 \tag{1.36}$$

Where, $\hat{o_k}$, and o_k are the desired and actual outputs, respectively. The inclusion of quantum rotation gates and the hybrid architecture enhances learning capabilities and adaptability of the model.

1.5.2.3 Quantum Dot Neural Network (QDNN)

It was proposed by Behrman et al. in 1996 [34]. A QDNN is a neural network model that utilizes quantum dots as computational units to represent neurons or synapses. Quantum dots are nanoscale semiconductor particles used for several experimental purposes. Quantum dots exhibit quantum mechanical properties, such as discrete energy levels, which can be used for information processing. Each quantum dot can represent a qubit, which is a linear combination of basis states $|0\rangle$ and $|1\rangle$. The activation function in a QDNN uses quantum mechanical operations, such as inner products or phase shifts. The output of a quantum dot neuron is given by the inner product between the input state $|\psi_i\rangle$ and weight state $|w_{ij}\rangle$. The pre-activation value h_j for the j-th quantum neuron can be formulated as:

$$h_j = \sum_{i=1}^{n} \langle \psi_i | w_{ij} \rangle \tag{1.37}$$

The quantum dot system can incorporate nonlinear activation functions $f(.)$, like Sigmoid or RELU when applied to pre-activation values:

$$y_j = f(h_j) \tag{1.38}$$

The final output is computed by combining the output values produced by the neurons in the hidden layer with another set of weights and applying an activation function. For m output neurons:

$$y_k = f\left(\sum_{j=1}^{p} w_{jk} h_j\right), k = 1, 2, ..., m \tag{1.39}$$

The learning process involves adjusting the quantum dot weights to minimize error, such as mean squared error. In QDNN, the states of the system evolve over time and are described using quantum mechanics equations.

1.5.2.4 Quantum Cellular Neural Network (QC_lNN)

QC_lNN combines the quantum computing principles with classical C_lNN [35]. C_lNNs are a class of nonlinear dynamic systems designed to solve complex tasks such as image processing and pattern recognition [36]. Similar to classical C_lNNs, QC_lNNs work on cells with local interactions. However, quantum mechanics adds superposition and entanglement, allowing them to represent information in a more detailed manner. In a network of M qubits, the state of the entire system in QC_lNN is represented as:

$$|\Phi\rangle = \sum_{i=0}^{2^M-1} d_k\,|k\rangle \tag{1.40}$$

Where, d_k are complex coefficients that satisfy the normalization condition $\sum |d_k|^2 = 1$. The evolution of QC_lNNs can also be described using a Hamiltonian $\mathcal{H}$, which is responsible for the time evolution of the quantum system:

$$i\hbar\frac{\partial}{\partial t}\,|\Phi(t)\rangle = \mathcal{H}\,|\Phi(t)\rangle \tag{1.41}$$

The Hamiltonian $\mathcal{H}$ for QC_lNNs includes terms representing local interactions and external potentials:

$$\mathcal{H} = \sum_i \mathcal{H}_i + \sum_{i,j} \mathcal{H}_{ij} \tag{1.42}$$

Where, $\mathcal{H}_i$ represents the self-Hamiltonian of a cell and $\mathcal{H}_{ij}$ represents the interaction between neighboring cells.

1.5.2.5 Qubit Neural Network

The concept of qubit neural network was introduced by Matsui et al. [37] in 2000. In their model, the firing and non-firing states of neurons were mapped to the quantum states $|0\rangle$ and $|1\rangle$, respectively, while an arbitrary neuron state existed as a superposition of these two states. This allowed the network to process multiple possibilities simultaneously, leveraging quantum features like superposition. The model introduced two key parameters: the phase parameter T, which controlled the relationships between neurons, and the reversal parameter G, which influenced individual neuron's behavior. The neuron states evolved over time through quantum operations based on inputs, with the phase and reversal parameters adjusting the network's behavior. The model aimed to improve neural network efficiency by leveraging quantum properties, providing a framework for training quantum-inspired networks and applying quantum principles to machine learning tasks.

In 2003, there was an effort to enhance the backpropagation algorithm by integrating quantum principles to facilitate the learning process within a qubit neural network [38]. The learning updates are defined by the following equations [29]:

$$\begin{aligned} h_l^{new} &= h_l^{old} - \eta\frac{\partial E_{total}}{\partial h_l} \\ k^{new} &= k^{old} - \eta\frac{\partial E_{total}}{\partial k} \\ d^{new} &= d^{old} - \eta\frac{\partial E_{total}}{\partial d} \end{aligned} \tag{1.43}$$

Where, h_l^{new} and h_l^{old} represent the updated and original values of the hidden layer activations, respectively; k^{new} and k^{old} are the updated and initial weights, respectively; d^{new} and d^{old} represent the modified and original adjustment terms; and η represents the learning parameter.

1.5.2.6 Quantum Competitive Neural Network (QCNN)

It was introduced by Zhou in 2010 [39]. It integrates the classical competitive neural network (CCNN) with principles from quantum mechanics.

- **Classical competitive neural network**:
  It consists of two distinct layers. In the first layer, the Euclidean distance between the input pattern $\hat{Y}$ and the stored patterns is calculated, aiming to identify the closest match. The second layer utilizes a competitive learning rule to find the most similar stored pattern to the input. The neuron closest to the input, based on the smallest Euclidean distance, is selected as the winner. Before processing, both the input pattern $\hat{Y}$ and the weight vectors for each layer $\hat{W}_l$ are normalized. This normalization is important for ensuring that the vectors are compared in a fair manner. Let the input pattern $\hat{Y} = [y_1, y_2, ..., y_t]$ be vectors with t elements. $\hat{Y}$ can be normalized as:

$$\hat{Y} = \frac{Y}{\|Y\|} \tag{1.44}$$

  Where, $\|Y\|$ is the Euclidean magnitude of the input vector, calculated as:

$$\|Y\| = \sqrt{\sum_{k=1}^{t} {Y_k}^2} \tag{1.45}$$

  Putting the value of $\|Y\|$ obtained in Eq. (1.34), the Eq. (1.33) now becomes:

$$\hat{Y} = \frac{Y}{\|Y\|} = \left[\frac{y_1}{\sum_{i=1}^{t} y_k^2}, \ldots, \frac{y_t}{\sum_{k=1}^{t} y_k^2}\right] \tag{1.46}$$

  Similarly, the weight vector $\hat{W}_l, l = 1, 2, ..., m$ is normalized as:

$$\hat{W}_l = \frac{W_l}{\|W_l\|} = \left[\frac{w_{1l}}{\sum_{i=1}^{t} w_{il}^2}, \ldots, \frac{w_{ml}}{\sum_{i=1}^{t} w_{il}^2}\right] \tag{1.47}$$

  Then the winning neuron is selected using *cosine* similarity. The cosine similarity between the input pattern $\hat{Y}$ and the weight vector $\hat{W}_l$ is given by:

$$cos\phi = \frac{\hat{W}_l.\hat{Y}}{\left\|\hat{W}_l\right\| \left\|\hat{Y}\right\|} \tag{1.48}$$

Where, $\hat{W}_l$ is the normalized weight vector of the l-the neuron, $\hat{W}_l.\hat{Y}$ is the dot product between the normalized weight vector and the normalized input vector. The neuron with maximum cosine similarity is called the winning neuron. Mathematically, it can be represented as:

$$l^* = arg \max_l cos\phi_l \tag{1.49}$$

The similarity between two patterns increases as the angle ϕ decreases.

- **QCNN**:
CCNN, QCNN has also two layers: input layer, and competitive layer. The main difference lies in the underlying principles they use for similarity comparison and neuron selection. In QCNN, the normalization process involves quantum states. The input quantum state $\left|\hat{Y}\right\rangle$ and the quantum weight vector $\left|\hat{W}_l\right\rangle$ are normalized similarly as in CCNN. However, they differ in similarity measures adopted. CCNN uses cosine similarity based on Euclidean distance between the input and weight vectors, whereas QCNN uses the *quantum cosine similarity* based on the quantum inner product:

$$cos\phi = \langle W_l|Y\rangle \tag{1.50}$$

Where, $\langle W_l|Y\rangle$ represents the quantum inner dot product between the quantum state of the stored pattern $|W_l\rangle$ and the quantum state of the input pattern $|Y\rangle$. The neuron with the maximum quantum dot product is selected as the winning neuron:

$$l^* = arg \max_l \langle W_l|Y\rangle \tag{1.51}$$

1.5.2.7 Quantum Associative Neural Network (QANN)

Zhou et al. [40] introduced the concept of QANN with non-linear search algorithm, integrating the principles of associative neural networks [41] with quantum computing. QANN operates in two main stages: the first stage involves storing the patterns, and the second stage involves pattern recalling which is similar to the classical associative NN. In QANN, pattern storage is achieved through quantum operations and quantum gates. The authors introduced a quantum binary decision diagram (QBDD) to represent stored patterns, utilizing the structure of a binary tree. The QBDD is constructed using the principle of quantum superposition, where quantum states encode multiple possibilities simultaneously. In the QBDD depicted in Fig. 1.3, nodes $n_a, n_b, ..., n_l$ represent nodes in the binary tree, with n_a acting as the root node. Quantum gates $a, b, ..., l$ are applied to transform states during the storage process. The states $|\phi_a\rangle, ..., |\phi_l\rangle$ are the quantum states, and $\left|\phi_g\right\rangle, |\phi_h\rangle, ..., |\phi_l\rangle$ are the desired stored quantum states encoding the patterns.

The pattern recall mechanism in QANN is similar to that of classical associative memory. The process retrieves the closest matching stored pattern when a partial or noisy input is provided. This is achieved through a nonlinear search algorithm that

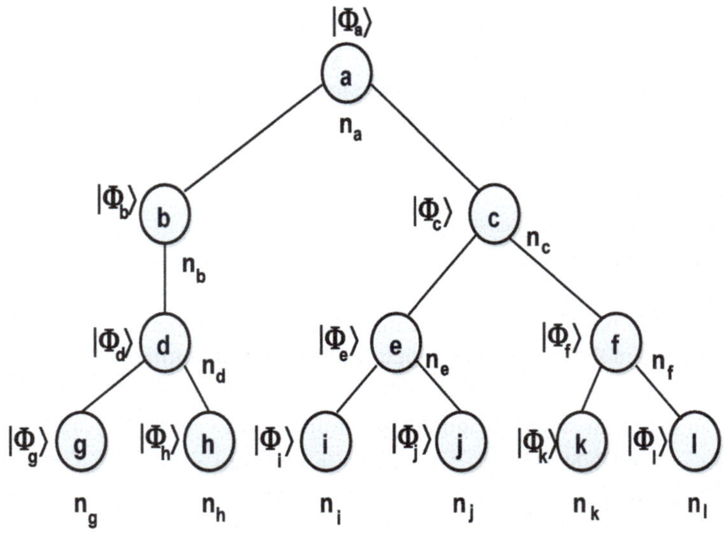

Fig. 1.3 Quantum binary decision diagram

efficiently navigates the quantum binary decision diagram to identify and measure the relevant quantum state corresponding to the stored pattern.

1.6 Conclusion and Future Work

The quantum enhanced bioinspired algorithms represent a transformative advancement in computational science. This chapter explored the strengths of both fields, highlighting how unique capabilities of quantum computing can enhance the efficiency and performance of bioinspired algorithms like genetic algorithms and neural networks. Despite the advancements in this field, challenges still remain, such as the requirement for scalable quantum networks, efficient quantum-classical hybrid approaches, and robust error correction techniques. In this chapter we focused on quantum computing applications on GA and NN. In future, we aim to explore the hybridization of quantum computing with other bioinspired algorithms like PSO, BBO and ACO. With the advancement of quantum technology, these techniques could solve real-world problems in areas like climate science, robotics, and system design, opening new opportunities for innovation. Furthermore, as quantum technology continues to evolve and becomes more readily available, we anticipate a rise in collaboration between quantum researchers and industry professionals to address the complex, large-scale challenges.

References

1. Darwish, A.: Bio-inspired computing: algorithms review, deep analysis, and the scope of applications. Future Comput. Inf. J. **3**(2), 231–246 (2018)
2. Hakemi, S., Houshmand, M., KheirKhah, E., Hosseini, S.A.: A review of recent advances in quantum-inspired metaheuristics. Evol. Intell. **17**(2), 627–642 (2024)
3. Dandan, H., Li, X., Liu, C., Liu, Z.-W.: Integrating environmental and economic considerations in charging station planning: an improved quantum genetic algorithm. Sustainability **16**(3), 1158 (2024)
4. Schuld, M., Sinayskiy, I., Petruccione, F.: The quest for a quantum neural network. Quantum Inf. Process. **13**, 2567–2586 (2014)
5. Han, K.-H., Kim, J.-H.: Quantum-inspired evolutionary algorithm for a class of combinatorial optimization. IEEE Trans. Evol. Comput. **6**(6), 580–593 (2002)
6. Samuels, G., Dutta, D., Mahon, P., Nikam, S.V.: The importance of bell states in quantum computing. In: *16th International Conference on Information Technology-New Generations (ITNG 2019)*, pp. 581–585. Springer, Berlin
7. Hou, Min, Yue, Wu.: New quantum private comparison using bell states. Entropy **26**(8), 682 (2024)
8. Williams, C.P., Williams, C.P.: Quantum gates. Explor. Quantum Comput. 51–122 (2011)
9. Myers, D.J., Sati, H., Schreiber, U.: Topological quantum gates in homotopy type theory. Commun. Math. Phys. **405**(7), 172 (2024)
10. Karthikeyan, S., Akila, M., Sumathi, D., Poongodi, T.: Quantum Machine Learning: A Modern Approach. CRC Press (2024)
11. Fan, X., Sayers, W., Zhang, S., Han, Z., Ren, L., Chizari, H.: Review and classification of bio-inspired algorithms and their applications. J. Bionic Eng. **17**, 611–631 (2020)
12. Lambora, A., Gupta, K., Chopra, K.: Genetic algorithm-a literature review. In: *2019 International Conference on Machine Learning, Big Data, Cloud and Parallel Computing (COMITCon)*, pp. 380–384. IEEE (2019)
13. Alhijawi, B., Awajan, A.: Genetic algorithms: theory, genetic operators, solutions, and applications. Evol. Intell. **17**(3), 1245–1256 (2024)
14. Awadallah, M.A., Makhadmeh, S.N., Al-Betar, M.A., Dalbah, L.M., Al-Redhaei, A., Kouka, S., Enshassi, O.S.: Multi-objective ant colony optimization. Arch. Comput. Methods Eng. 1–43 (2024)
15. Dorigo, M., Stützle, T.: Ant Colony Optimization: Overview and Recent Advances. Springer, Berlin (2019)
16. Akbari, R., Hedayatzadeh, R., Ziarati, K., Hassanizadeh, B.: A multi-objective artificial bee colony algorithm. Swarm Evol. Comput. **2**, 39–52 (2012)
17. Hedayatzadeh, R., Hasanizadeh, B., Akbari, R., Ziarati, K.: A multi-objective artificial bee colony for optimizing multi-objective problems. In: *2010 3rd International Conference on Advanced Computer Theory and Engineering (ICACTE)*, vol. 5, pp. V5–277. IEEE (2010)
18. Simon, D.: Biogeography-based optimization. IEEE Trans. Evol. Comput. **12**(6), 702–713 (2008)
19. Ma, H., Simon, D., Siarry, P., Yang, Z., Fei, M.: Biogeography-based optimization: a 10-year review. IEEE Trans. Emerg. Top. Comput. Intell. **1**(5), 391–407 (2017)
20. SaiToh, A., Rahimi, R., Nakahara, M.: A quantum genetic algorithm with quantum crossover and mutation operations. Quantum Inf. Process. **13**, 737–755 (2014)
21. Talbi, H., Draa, A., Batouche, M.: A new quantum-inspired genetic algorithm for solving the travelling salesman problem. In: 2004 IEEE International Conference on Industrial Technology, 2004. IEEE ICIT'04, vol. 3, pp. 1192–1197. IEEE (2004)
22. Lahoz-Beltra, R.: Quantum genetic algorithms for computer scientists. Computers **5**(4), 24 (2016)
23. Mohammed, A.M., Elhefnawy, N.A., El-Sherbiny, M.M., Hadhoud, M.M.: Quantum crossover based quantum genetic algorithm for solving non-linear programming. In: 2012 8th International Conference on Informatics and Systems (INFOS), pp. BIO–145. IEEE (2012)

24. Xu, Z., Shang, W., Kim, S., Bobbitt, A., Lee, E., Luo, T.: Quantum-inspired genetic algorithm for designing planar multilayer photonic structure. NPJ Comput. Mater. **10**(1), 257 (2024)
25. da Silveira, L.R., Tanscheit, R., Vellasco, M.M.B.R.: Quantum inspired evolutionary algorithm for ordering problems. Exp. Syst. Appl. **67**, 71–83 (2017)
26. Wang, L.-Y., Zhang, J., Li, H.: An improved genetic algorithm for TSP. In: 2007 International Conference on Machine Learning and Cybernetics, vol. 2, pp. 925–928. IEEE (2007)
27. Abbas, A., Sutter, D., Zoufal, C., Lucchi, A., Figalli, A., Woerner, S.: The power of quantum neural networks. Nat. Comput. Sci. **1**(6), 403–409 (2021)
28. Zhou, M.-G., Liu, Z.-P., Yin, H.-L., Li, C.-L., Tong-Kai, X., Chen, Z.-B.: Quantum neural network for quantum neural computing. Research **6**, 0134 (2023)
29. Jeswal, S.K., Chakraverty, S.: Recent developments and applications in quantum neural network: a review. Arch. Comput. Methods Eng. **26**(4), 793–807 (2019)
30. Zhou, R., Ding, Q.: Quantum MP neural network. Int. J. Theor. Phys. **46**, 3209–3215 (2007)
31. McCulloch, W.S., Pitts, W.: A logical calculus of the ideas immanent in nervous activity. Bull. Math. Biophys. **5**, 115–133 (1943)
32. Menneer, T., Narayanan, A.: Quantum-inspired neural networks. Proc. Neural Inf. Process. Syst. **95**, 27–30 (1995)
33. Shang, F., et al.: Quantum-inspired neural network with quantum weights and real weights. Open J. Appl. Sci. **5**(10), 609 (2015)
34. Behrman, E.C., Niemel, J., Steck, J.E., Skinner, S.R.: A quantum dot neural network. In: Proceedings of the 4th Workshop on Physics of Computation, pp. 22–24 (1996)
35. Tóth, G., Lent, C.S., Tougaw, P.D., Brazhnik, Y., Weng, W., Porod, W., Liu, R.-W., Huang, Y.-F.: Quantum cellular neural networks. Superlattices Microstruct. **20**(4), 473–478 (1996)
36. Chua, L.O., Yang, L.: Cellular neural networks: applications. IEEE Trans. Circuits Syst. **35**(10), 1273–1290 (1988)
37. Matsui, N., Takai, M., Nishimura, H.: A network model based on qubitlike neuron corresponding to quantum circuit. Electron. Commun. Jpn. (Part III: Fundamental Electronic Science) **83**(10), 67–73 (2000)
38. Kouda, N., Matsui, N., Nishimura, H.: A multilayered feed-forward network based on qubit neuron model. Syst. Comput. Jpn. **35**(13), 43–51 (2004)
39. Zhou, R.: Quantum competitive neural network. Int. J. Theor. Phys. **49**, 110–119 (2010)
40. Zhou, R., Wang, H., Qian, W., Shi, Y.: Quantum associative neural network with nonlinear search algorithm. Int. J. Theor. Phys. **51**(3), 705–723 (2012)
41. Peruš, M.: Neural networks as a basis for quantum associative networks. Neural Netw. World **10**(6), 1001–1013 (2000)

Chapter 2
Decoding Market Dynamics: Variational Quantum Circuit in Stock Prediction

Bikash Chandra Sahoo, Sandeep Kumar Satapathy, Sung-Bae Cho, and Shruti Mishra

Abstract Financial markets known for their everchanging and intricate nature have been a huge topic in stock market prediction research. Traditional methods typically use regular computer-based learning to figure out patterns in past data. The game-changing impact of quantum machine learning on predicting stock markets is significant. Quantum computing with its principles of superposition and entanglement have the ability to handle complex calculations and analyze large datasets at once shows promising results. In this work, a quantum variational circuit is used which is a classifier and is designed to optimize parameters and find the best solution to a problem which in our case is to learn pattern and trends. The circuit was incorporated using different optimizers like "Real Amplitude" and "efficientSU2" to obtain the best possible results. With this we achieveda very high accuracy measured on the basis of quantum variational score when applied to a cryptocurrency-based stock dataset. However, despite the outstanding results it takes more time because of unavailability of widespread access to commercial quantum computers but it certainly holds potential to equally predict as the trivial methods.

Keywords Stock · Quantum machine learning · Variational quantum circuit · Optimizers

B. C. Sahoo
School of Computer Science and Engineering, Vellore Institute of Technology, Chennai, Tamil Nadu, India
e-mail: bikashchandra.sahoo2022@vitstudent.ac.in

S. K. Satapathy (✉) · S. Mishra
Centre for Advanced Data Science, Vellore Institute of Technology, Chennai, Tamil Nadu, India
e-mail: sandeepkumar04@gmail.com

S.-B. Cho
Department of Computer Science, Yonsei University, Seoul, South Korea
e-mail: sbcho@yonsei.ac.kr

S. Dehuri et al. (eds.), *Advances in Quantum Inspired Artificial Intelligence*, Intelligent Systems Reference Library 274, https://doi.org/10.1007/978-3-031-89905-8_2

2.1 Introduction

The study of the financial market has become a subject of interest to investors and analysts mainly because every aspect, combination or component is used to understand and predict as well as being influenced by substantial changes [1, 2]. However, the vast and convoluted datasets of these markets have made employing classic machine learning algorithms to be difficult tasks. One such path of exploration has been the quantum machine learning (QML) that is evolving as a disruptive way to improve computation capabilities in machine learning using principles from quantum mechanics [3].

Quantum algorithms are expected to lead a revolution by utilizing superposition and entanglement properties of quantum computers, machine learning, especially in solving computationally expensive problems like NP hard tasks [3]. A more detailed analysis of the issue regarding quantum computing impact on data mining and machine learning is finding that even though specific patterns such as pattern recognition [4] have benefitted from development in quantum-based algorithms.

The work by Kopczyk [5] strives to make quantum machine learning algorithms accessible to data scientists by avoiding overly complex mathematical formulations. The step-by-step explanations were provided including the quantum PCA algorithm and the enumeration of necessary algorithms for different applications contribute to the comprehensibility of quantum machine learning [6].

As quantum computers become more accessible, the need to train a new cadre of quantum programmers arises. A survey examines 20 different quantum algorithms, discussing their applications on IBM's quantum computer and highlighting the differences between simulations and actual hardware runs. This emphasizes the ongoing evolution and improvement of quantum computing technology in terms of qubit count, quality and connectivity.

Incorporating QML also involves exploring various quantum computing frameworks, models and techniques [7]. Notable libraries for implementing quantum algorithms such as Qiskit, Pennylane and Tensorflow-quantum are widely used. The Noisy Intermediate Scale Quantum (NISQ) approach aims to effectively demonstrate the supremacy of quantum computing. Review papers delve into fundamental QML algorithms like Quantum Support Vector Machine (QSVM), Quantum Principal Component Analysis (QPCA), Quantum K-Nearest Neighbor (Q-KNN), Quantum Kernel Matrix, and Grover's algorithm [8].

Despite the promising merits of QML, it comes with its fair share of challenges [9]. Current quantum computers face limitations in terms of qubit count, coherence duration and the likelihood of errors. Adapting regular data to a quantum form and handling quantum data efficiently pose challenges. Error correction and noise reduction are critical for making quantum machine learning reliable. Quantum Neural Network (QNN) introduces a quantum–classical hybrid model but is relatively complex. Developing effective strategies for combining the strengths of both quantum and classical approaches remains an ongoing challenge, especially considering the development phase of many quantum libraries and the deprecation of certain functions [10].

While quantum machine learning has shown clear benefits in specific learning tasks such as those designed for Shor's algorithms, efforts are expanding to create a broader range of learning tasks that utilize regular data. This showcases the general strengths of quantum computing beyond the scope of Shor's capabilities [11]. Tech giants are already taking strides to harness the potential of quantum computing in various fields, with promising results in finance, medicine, and beyond. However, the full realization of quantum computing's power necessitates better-quality quantum hardware as the current hardware lacks the required quality, speed and size.

The ongoing evolution of technology and data underscores the importance of existing systems adapting and evolving. Thorough testing and validation are imperative to ensure that quantum machine learning is practically useful across different fields considering its current status as a technology in development. As the landscape continues to evolve, robust resources and system capabilities become essential to handle the increasing complexity and potential failures in the quantum computing realm.

2.2 Literature Survey

The stock market, a dynamic domain has captivated analysts and researchers aiming to provide accurate analyses for investors [1]. In one research by Kobayashi et al. [12], Quantum Neural Network (QNN) was employed, serving as a hybrid of classical and quantum networks. The tensor network, facilitating mathematical and pictorial representation, showcased a comparative analysis between classical and quantum models. Despite tensor network outperforming the classical model, it introduced complexity risks [12]. Mahajan's work [13] also utilized QNN, incorporating quantum neurons in hidden layers and classical neurons in output layers. It validated the Efficient Market Hypothesis, comparing the model to a regular computer model with the same information parts. Notably, increasing the information input improved the computer's prediction accuracy, with QNN exhibiting similar accuracy as classical models, and a slight advantage over CNN in terms of speed [13]. Cao et al. [14] conducted a comparative analysis between Linear Layered enhanced Quantum Long Short-Term Memory (LL-QLSTM) and QLSTM. LL-QLSTM outperformed QLSTM due to its adaptability to data patterns and an optimized circuit leading to improved convergence [14]. Naman et al. [15] employed Quantum Support Vector Machine (QSVM) for binary classification in finance-related data. QSVM generated a Quantum Kernel Machine, crucial for extracting important details from financial data. The study compared results to the classical SVM, which significantly outperformed QSVM. This indicated that quantum computers might not always surpass classical methods, emphasizing the need for a careful and articulated approach [15]. Alaminos et al. [16] explored stock market crashes with a comparative analysis between Support

Vector Regressive based Quantum BAT Algorithm (SVRQBA), Quantum Boltzmann Machine (QBM), and QNN. QBM demonstrated the highest accuracy of 94%, effectively handling quantum states' potential noise [16]. Paquet et al. [17] introduced a hybrid quantum neural network named Quantum Leap, using an encoder to convert financial time series data into coherent density matrices. The system demonstrated precision and effectiveness in both regression and extrapolation scenarios [17]. Liu et al. [18] implemented a Doubled Chain Quantum Algorithm (DCQA) in a quantum artificial neural network, validated on six stock markets for closing price prediction. The simulation results indicated the feasibility and effectiveness of the proposed algorithm [18]. Dimoska et al. [19] utilized Parametrized Quantum Circuit (PQC) for time series data, outperforming BiLSTM when facing significant variations. PQC's ability to navigate quantum states while determining data relations contributed to its success [19]. Marco et al. [20] explored various quantum algorithms based on application needs, identifying seven machine learning tasks achievable with different quantum algorithms [20]. Alaminos et al. [21] employed Deep Neural Decision Tree (DNDT), a quantum-based model that outperformed classical neural networks. DNDT combined quantum mechanics with decision trees, achieving superior results [21]. Adharsh et al. [22] monitored fractals in stock market graphical analysis using Chaos theory. Seven out of nine tested fractals accurately predicted movement, achieving an accuracy of 78% [22]. Ahmed et al. [23] implemented Time delay Additive Evolutionary Prediction method (TAEP) for time series data, demonstrating robustness and outperforming previous findings [23]. Guo et al. [24] utilized Continuous Variational Quantum Algorithm (CVQA) on time series data, showcasing its superiority over VQA in learning implicit dynamic patterns. Optimization techniques were explored for enhanced performance [24]. Chimprang et al. [25] conducted a comparative analysis on Quantum Particle Swarm Optimization (QPSO), Quantum Artificial Bee Colony (QABC), and Quantum Fruit Fly Optimization Algorithm (QFOA). QPSO showed promising results for stock index forecasting, albeit with high complexity [25]. Wang et al. [26] introduced Primary Ensemble Empirical Mode Decomposition (PEEMD) with QNN for forecasting stock index. PEEMD achieved the highest accuracy, emphasizing the effectiveness of combining PEEMD, QNN, and Back Propagation [26]. Huggins et al. [27] focused on image classification using TensorFlow-quantum. The model, utilizing tensor circuits, demonstrated the integration of quantum computing with machine learning for tasks like handwriting recognition [27]. Takaki et al. [28] introduced a method using parametrized quantum circuit for temporal data, mimicking Recurrent Neural Network. The approach showcased how quantum computers can handle complex patterns [28]. Killoran et al. [29] structured a continuous variable quantum circuit using Strawberry fields library. The Quantum Neural Network (QNN) demonstrated the unique ability to handle complicated tasks while maintaining clarity [29] (Table 2.1).

Table 2.1 Previous research analysis

S. no	Authors	Dataset	Model used	Input	Response	Findings/Analysis
1	Mahajan [13]	Bharati, Chambel fertilizers, Maruti and MRF stock of 7 years	QNN composed of 30 quantum hidden neurons and 30 classical output neurons	Date, Open, High and Low prices with Closing price as the target variable	Classified and predicted the closing price by generating cost function	QNN has slightly performed better than CNN which is around 4%. Feature availed by QNN i.e., quantum parallelism helped in reducing the processing time. Extra memory would have fulfilled the greater chance of covering an extensive stock
2	Dimitrios and Dimoska [19]	AAPL stock, BTC-USD stock	Parametrized quantum circuit as QNN. It consisted of 6 layers of gates incorporated with controlled parametrized gate	Time series data and the closing price	Forward propagation of network was quantum whereas the backpropagation was classical. Temporal signals with SNR value were fetched for benchmarking	On comparison with BiLSTM, PQC outperformed when there is a huge variation in noise. The propagations could be made completely quantum if the continuous optimization algorithm becomes robust
3	Kobayashi et al. [12]	TOPIX500 index of Japanese stock market	Two neural networks governed by ReLU consisting of 3 and 4 layers respectively, One PQC based circuit and a Tensor network	10 features which could be generalized as Value, Quality, Momentum, Size and Market	The models felicitate empirical backtesting consisting of excess return, tracking error and information ratio	Quantum circuit outperforms classical linear model due to its ability to learn non-linear relations. The tensor network achieved the highest performance even though it was fed with less parameters

(continued)

Table 2.1 (continued)

S. no	Authors	Dataset	Model used	Input	Response	Findings/Analysis
4	Cao et al. [14]	ECX of EU Emission trading system	Quantum LSTM augmented with linear layers, featuring additional linear layers both preceding and succeeding the variational quantum circuit	Time series data consisting of Historical emissions, Emission by sector and aviation, Surrendered units and entities by size	Deviation in the form of errors by using different combination of layers fetched the differences between QLSTM and L-QLSTM	L-QLSTM outperformed QLSTM by achieving the least error in both training and testing set. Layers used in VQC helped reduce qubit usage thus making the learning process better and hence required lesser features for prediction
5	Srivastava et al. [15]	Stock prices of Honeywell, Johnson and Johnson, Apple and Visa	Quantum annealing for feature selection and QSVM for classification	Moving average, Relative strength index, Moving average convergence divergence, stochastic oscillator and Aroon indicators	The inbuilt quantum kernel in QSVM fetches the correlation among the features. Accuracy using four different entanglements were found out	The quantum annealing helped in achieving the most dominant features precisely. There was no significant advantage of QSVM over its classical counterpart i.e., SVM in terms of accuracy
6	Alaminos et al. [16]	Stock data from FRED database, World bank open data and IMF	Quantum Boltzmann Machine	32 independent variables classified as domestic prices, Open economy, commodities and financial and domestic real macro	Cost function to identify the fractals and henceforth spot the crashes	QBM achieved the highest accuracy of 96.22% when compared to SVRQBA. The model offers crucial new factors that agents can use to forecast declines in economic activity. As a result, the insights from our model provide vital information for policymakers

(continued)

Table 2.1 (continued)

S. no	Authors	Dataset	Model used	Input	Response	Findings/Analysis
7	Paquet and Soleymani [17]	24 stocks of companies namely Apple, Amazon, Cisco etc.	Encoder and a Quantum neural network with 5 hidden layers	Time series data consisting of stock prices, moving average, volume etc.	Sequence of density matrices via encoder and error calculation via prediction	No sign of overfitting while executing the model. The model shows the highest accuracy while comparing with two classical model on the basis of p-value performed using ARIMA method
8	Liu and Ma [18]	BSE Sensex, HSI, SSE, Russell 2000, and TAIEX	Quantum Elman Neural Network (QENN) and tuning using double chain quantum genetic algorithm	Time series data with closing price as the target variable	Temporal dependencies that captured volatility, non- linearity for the prediction of closing price	The model is highly effective in capturing the dynamic insights of stock. The tuning process using genetic algorithm gives the model an upper hand for increasing the learning rate
9	Alaminos et al. [21]	GDP of 70 countries from IMF International Financial Statistics and world bank	Support vector regression quantum BAT algorithm, QBM, QNN and other classical neural network	24 independent variables classified as financial variables and macroeconomic variables	Classification of GDP based on Emerging, Developed and Global country. Accuracy of algorithm for comparing quantum and classical methods	The deep neural decision tree (DNDT) achieved the highest precision. Accuracy range of algorithms varies from 93 to 98.9%. Found dominant features which could contribute more in GDP prediction. The model serves as a benchmark for formulating macroeconomic policy and enhancing decision-making processes
10	Bulusu et al. [22]	Stocks of 7 international banks from Yahoo Finance	Custom model based on chaos principle	Time series data consisting of price movement and sentiments retrieved from twitter and other news sources	Fractal analysis and sentiment analysis which were integrated to obtain prediction	The model achieved an accuracy of 78%. Out of 9 fractals, 7 fractals were predicted accurately showing the promising results. With the added advantage of quantum computing the model could achieve better performance in terms of runtime

(continued)

Table 2.1 (continued)

S. no	Authors	Dataset	Model used	Input	Response	Findings/Analysis
11	Weng et al. [24]	Climate time series data from weather underground API	Continuous variational quantum algorithm (CVQA) based circuit	Mean temperature, humidity, wind speed and mean pressure	Measurement values in the form of time. Obtained gradient using adjoint sensitivity method	On comparison with normal variational circuit, CVQA fetched the pattern in a better manner. The CVQA achieved the same accuracy as the QLSTM. The model was highly feasible for climate time series data
12	Takaki et al. [28]	Temporal dataset	Quantum recurrent neural network (QRNN)	Time stamp, time series data, event sequences, temporal features	Density matrix and expected values of a set of commuting observables for prediction	Identified the non-linearity in the temporal data. To advance to the next time step, the model needs to execute the entire QRNN up to a specific time point, t, to secure a prediction that meets certain accuracy criteria
13	Wang et al. [26]	Six stock data comprising of CSI 300, SSE 50, SSE 180, SZSE Index, SSEC Index, GEM Index	Primary ensemble empirical mode decomposition (PEEMD) QNN	Decomposed time series data equivalent intrinsic mode functions and low frequency eigen functions	Sequence of modal decomposition with varied frequencies to achieve predictive outcomes	Achieved different predictions based on different intrinsic modal functions due to noise and unused high frequency component. On comparison with EMD-QNN and QNN, PEEMD-QNN resulted in the lowest error and hence a promising model for trend judgement and error control
14	Huggins et al. [27]	MNIST	Discriminative and generative tensor network	Handwritten images with labels, image size and pixel values	Matrix product state and output based on classification. Accuracy based on pairwise classifier	Regarding noise impacts in tensor networks suggests that an error affecting a qubit only disrupts the information originating from the input patch associated with the qubit's past "causal cone". Digit 1 was identified with the highest accuracy while digit 4 with the lowest

2.3 Methodology

The research and design efforts are directed towards the prediction of stock market trends through the utilization of quantum computing capabilities, aiming to provide valuable insights for investors. The dataset utilized includes attributes such as Date, Open, High, Low, Close, Market cap and Volume which constitute a standard dataset available from platforms like Kaggle and other data-providing sites. During the preprocessing phase tasks like data imputation for missing values, data scaling, feature selection based on the correlation matrix and quantum encoding were conducted to transform classical data into quantum states, ensuring optimal input for the subsequent quantum circuit or simulator. The resulting quantum state equivalent to the input data is then fed into the quantum circuit driven by the algorithm. To facilitate classification, a new column named "price direction" was introduced indicating whether the closing price exceeded the opening price or vice versa. By choosing price direction as the target variable the data was trained using variational quantum circuit which is a type of classifier. Performance analysis involved metrics such as accuracy, F-1 score, and ROC with the results illustrating the cost function across iterations for various sets of optimizers and feature combinations.

2.3.1 Dataset

The dataset is a cryptocurrency based standard time series data consisting of Open, Low, High, Close etc. as the attributes. Itis available in csv format and could be fetched from either Kaggle or Quandl. Itconsists of 632,219 instances and eight features where the different digital currency was annotated using different code or symbols. It consists of three years of data ranging from 2015 to 2018. There are features like market cap which has some missing values. The features were selected based on the correlation matrix and the prominent features were fed for further processing.

Figure 2.1 represents the feature pair-wise in order to portray the observable correlation among the attributes extracted from feature selection. As it can be observed from the above that the price direction i.e., the up and down prices are going hand in hand. So, classification of it need to be found out by drawing optimal parameters which would help in bifurcating the two different classes.

2.3.2 Data Pre-processing and Transformation

Following the initial pre-processing steps applied to classical data such as handling missing values and selecting features the numerical values undergo transformation into quantum bits (qubits) through a quantum encoder [30, 31]. These values can

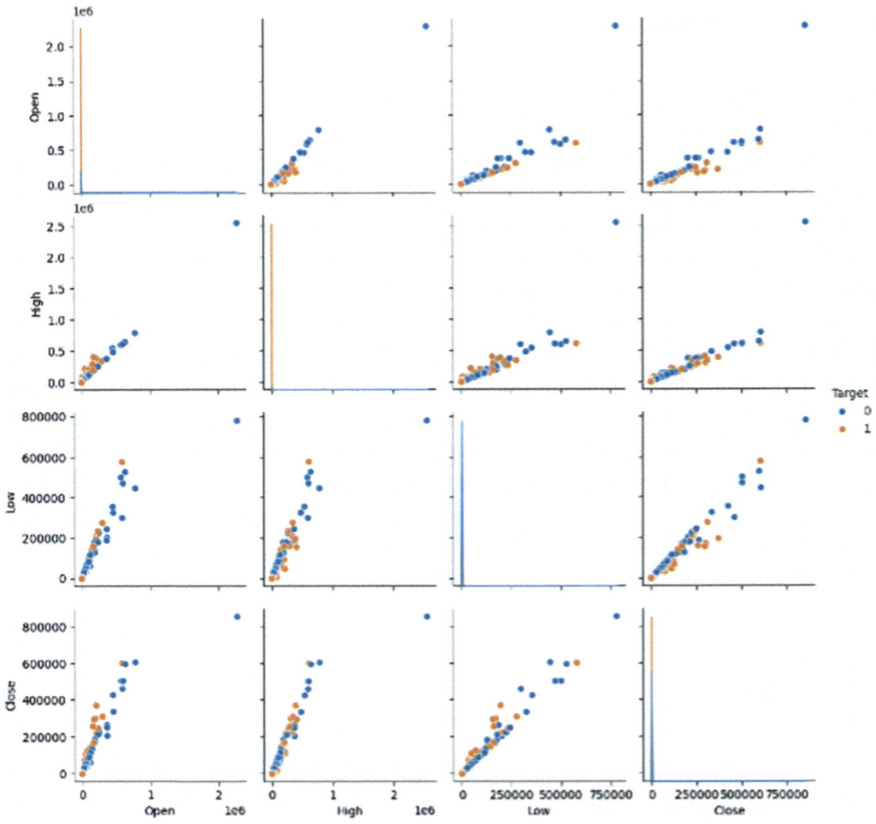

Fig. 2.1 Feature distribution

be symbolized using ket notation (|ψ⟩) to articulate the quantum state. In our case the qubits were represented by a combination of both real and complex values. To proceed with the subsequent stages of the process the quantum data must acquire the fundamental aspects of quantum mechanics specifically superposition and entanglement.

$$|\psi\rangle = \alpha|0\rangle + \beta|1\rangle \tag{2.1}$$

where, α and β are complex numbers.

Equation (2.1) represents the superposition of qubits. For the superposition, Hadamard gate was used which could be used either independently or incorporated with encoder circuits. The model uses the concept of feature map which is a type of encoder and has inbuilt Hadamard gate. The feature map encoder helpsin establishing correlation in such a manner that the qubits could remain in two different states simultaneously or could get entangled.

$$|\psi\rangle = \frac{1}{\sqrt{2}}(|00\rangle + |11\rangle) \tag{2.2}$$

Equation (2.2) shows the entanglement of quantum state. In order to achieve quantum entanglement.

Here, the qubits are entangled in the state where both are in a superposition of $|0\rangle$ and $|1\rangle$.

2.4 Model Overview

The proposed framework is implemented utilizing functionalities provided by the Quantum Machine Learning (QML) library. Some functions within QML are deprecated owing to its ongoing development phase but we have the flexibility to incorporate functions from diverse libraries that align with our specific objectives. The model employs a feature map encoder for quantum encoding addressing superposition and entanglement aspects. The variational quantum circuit guided by an optimizer, processes the data through a designated cost function. Enhancements are sought for the cost function which can be achieved by reducing features or employing alternative optimizers capable of minimizing the cost. The model training duration is prolonged due to reliance on traditional systems rather than quantum ones. Additionally, the model undergoes comparison with its classical algorithm counterpart to validate its accuracy.

2.4.1 Model Architecture

Figure 2.2 below showsfour basic phases of model architecture namely preprocessing, encoding, simulation and analysis. The classical data after going through the pre-processing stage is first morphed into quantum state via encoder which takes place in second phase. This encoder could be of any type based on desired task. The encoder is incorporated by gates like Hadamard and rotation which will help us attaining basic principles of quantum mechanics.

The quantum data will then be fed to the third phase i.e., simulation which will be composed of quantum circuits driven by quantum algorithm [32, 33]. Since we used variational quantum circuit it simply finds the optimal parameters. The optimal parameters are found out in the training process and hence accompany cost function during its iteration. The cost function needs to be taken care of especially in case of variational circuits as they are composed of parametric gates which involves in finding optimal parameters and hence could result in unoptimized output and long training time. To improve the cost function the features were further minimized using PCA and undergone simulation accompanied by different sets of optimizers to attain best possible results [34, 35].

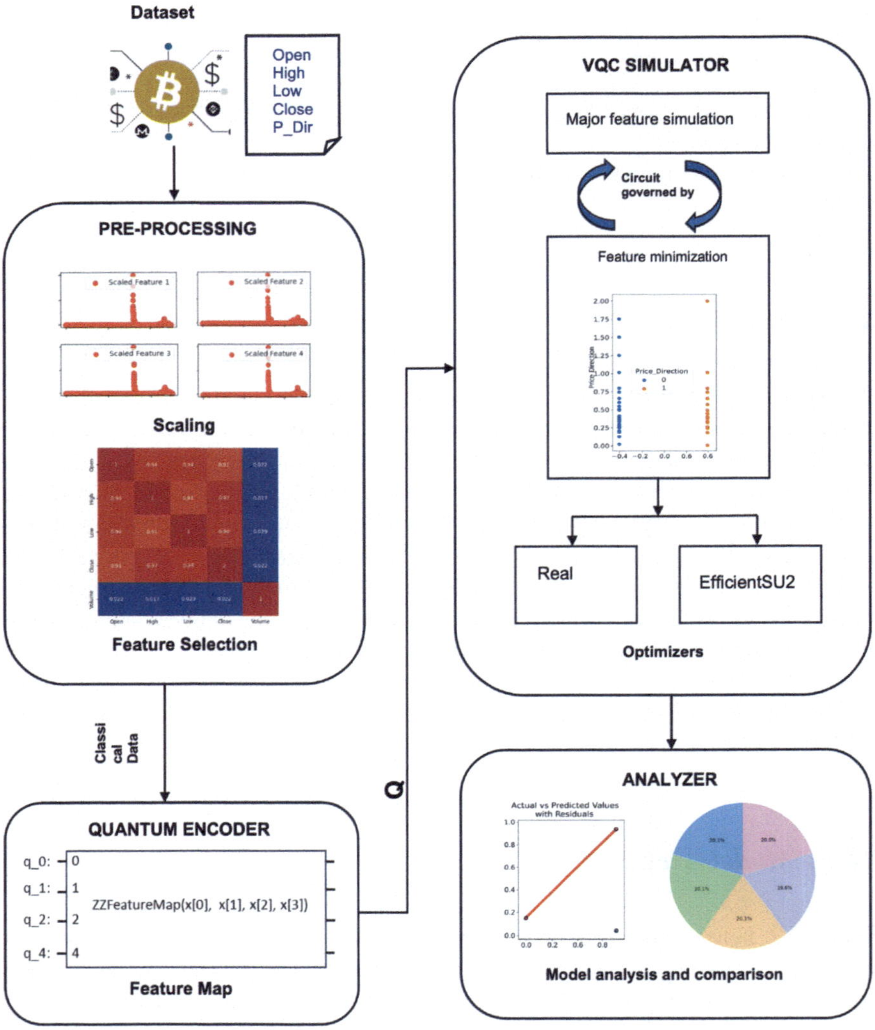

Fig. 2.2 Basic architecture of model

Finally, in the final phase the output is monitored based on the training and testing score. The performance metrics involved in a classification were also calculated to compare the superiority of algorithm in different conditions.

2.4.2 Model Design

Figure 2.3 represents the flow of implementation of variational quantum circuit. First the dataset was imported. The dataset was then pre-processed for any missing values, NaN and highly precise values. The missing values was filled with adjacent values while high precise values were scaled. On the basis of opening price and closing price a new column was generated annotating the price direction (0 for "price fall" and 1 for "price rise"). In the next phase the processed data was fed for selecting features which will act as the number of qubits for the simulator and will help in classification. After the feature selection the dataset was then split into 80–20 where 80 is the training set and 20 is testing set. The data was then fed to the feature map encoder which is consist of Hadamard gate, rotation gate parameterized by the classical feature, Pauli-X gate and Controlled-Z (CZ gate). The Hadamard gate performs the superposition while the CZ gate entangles the quantum state by performing a conditional phase shift on the target qubit based on the state of the control qubit. If the control qubit is in the state $|1\rangle$, it applies a phase shift of π (180°). The simulator is driven by specific algorithm which is variational quantum classification in our case. The training involves multiple iterations which will update the parameters for finding optimal one and hence attaining the best possible cost function. The same process will be carried out with other conditions such as decreasing features using PCA and passing other optimizers to compensate the cost function. Finally, the output will be validated based on training and testing score along with other performance metrics involved in classification.

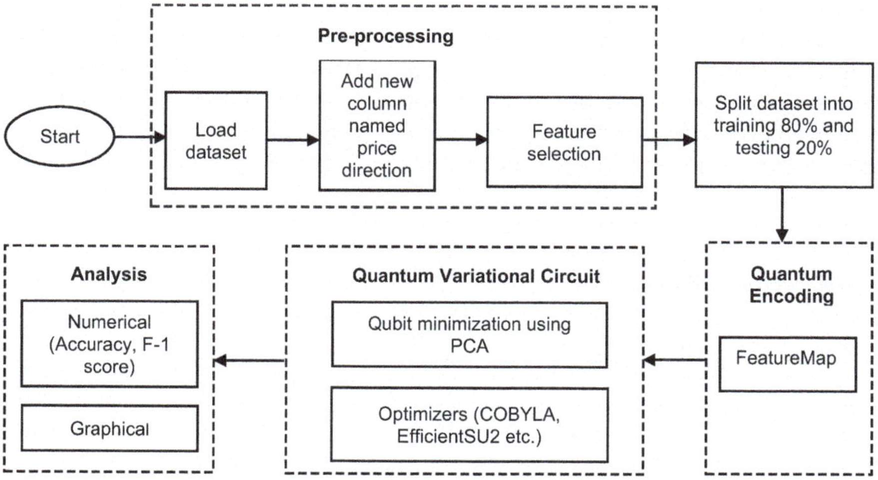

Fig. 2.3 Work flow model for quantum linear regression

2.4.3 Quantum Circuit Involved

Figure 2.4 illustrates the feature map encoder comprising multiple gates performing quantum encoding while applying quantum mechanical principles. The feature map quantum encoder transforms classical information into quantum systems typically utilizing qubits to implement superposition and leverage quantum mechanics. The ZZ interaction, a quantum gate in this context introduces a phase shift contingent on the qubit's state. The figure implies that the gate introduces a π radian phase shift when necessary. In the realm of feature map encoding this interaction is employed to represent relationships between input features.

The QML function facilitates the creation of a quantum circuit where quantum gates and operations transform regular information (such as numbers) into a unique quantum form using ZZ interactions [36, 37]. The qubits within the circuit are adjusted to discern connections between features and thus updating it to find the optimal parameters [38, 39]. The resulting quantum state is then employed for various tasks within quantum machine learning. The circuit's appearance and specifics depend on the information under consideration and the number of features used in the training process.

Figure 2.5 above is illustrating the Real Amplitudes ansatz which is a parameterized quantum circuit and contains adjustable gates. The parameters are tuned during the training to minimize the cost function [40, 41]. Initially, we are passing five features including the target variable and used three reps (number of repetitions) which will specify how many times it will repeat the alternating layers [42]. Based on the reps the number of parameters required to be adjusted could be visualized in the above given circuit.

Figure 2.6a, b above represents the two different optimizers used in conjunction with COBYLA (Constrained optimization BY Linear Approximation) optimizer fed with only two qubits. Similar to the previous circuit discussed, both will help in adjusting the parameters of gates in the quantum circuit to find the optimal configuration and henceforth attaining the best possible and minimal cost function.

2.4.4 Algorithm

1. Import necessary packages and libraries

2. Import dataset

3. Add Price direction column

3. Feature selection

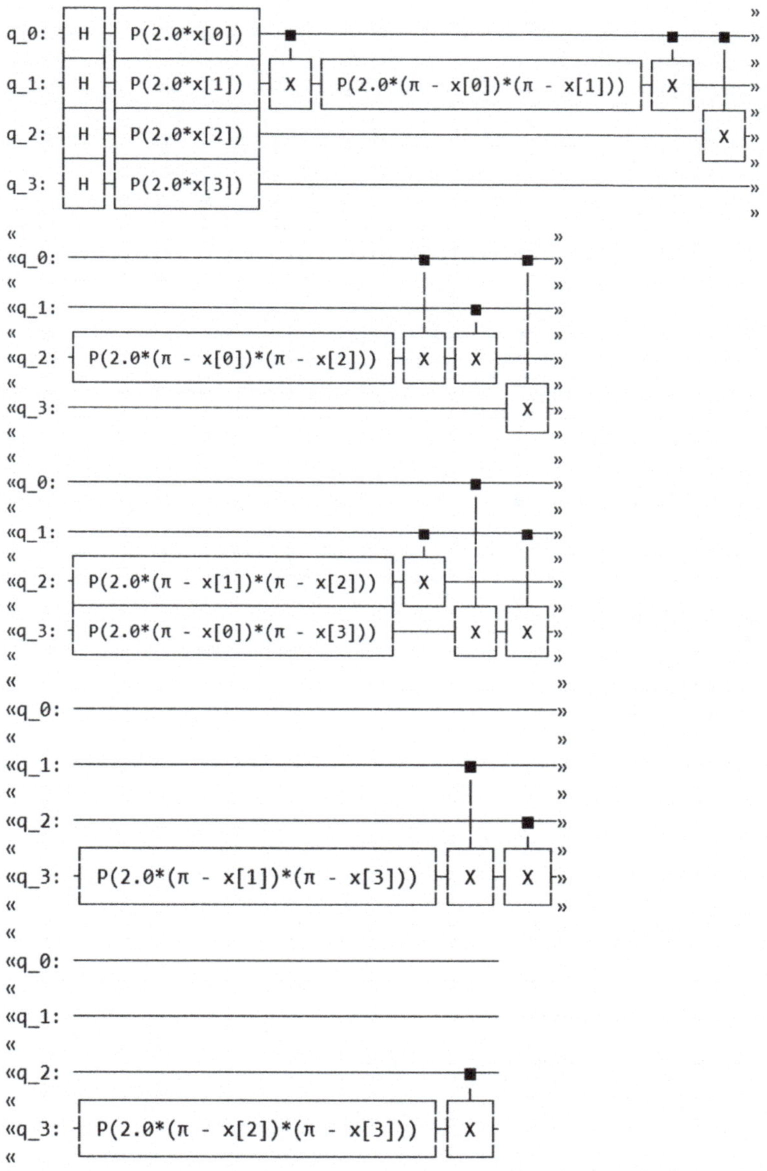

Fig. 2.4 Feature map quantum encoder

4. X-> input data

y-> target value

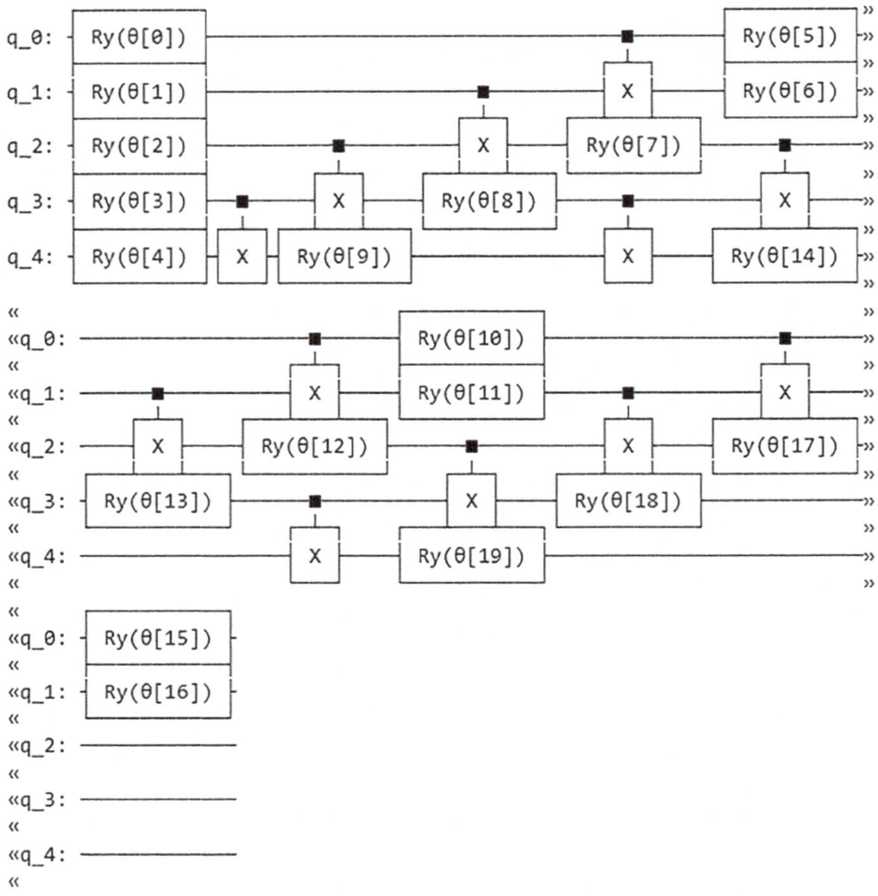

Fig. 2.5 Real amplitude optimizer with four qubits

5. Data encoding in quantum state

quantum_state = quantum_state_function (X, y)

6. Establish quantum circuit

7. Variational quantum circuit

quantum_process = quantum_algorithm(quantum_state)

8. Decrease qubit using PCA

9. Pass different optimizer

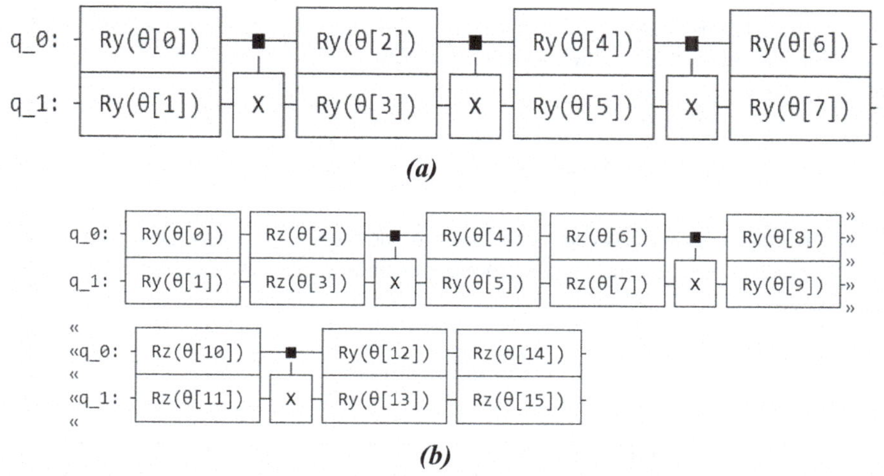

Fig. 2.6 **a** Real amplitude optimizer with two qubits. **b** EfficientSU2 optimizer with two qubits

10. Classical data retrieval (optional)

11. Statistical and performance metrics analysis

12. Graphical analysis

2.4.5 Mathematical Formulation

A. Hadamard operator

$$H = \frac{1}{\sqrt{2}}\begin{bmatrix} 1 & 1 \\ 1 & -1 \end{bmatrix}$$

It creates superposition by putting qubits in an equal probability combination of $|0\rangle$ and $|1\rangle$. For example, $H(|0\rangle) = \frac{1}{\sqrt{2}}(|0\rangle + |1\rangle)$.

B. Controlled Z gate (CZ)

$$CZ = \begin{bmatrix} 1 & 0 & 0 & 0 \\ 0 & 1 & 0 & 0 \\ 0 & 0 & 0 & 1 \\ 0 & 0 & 1 & -1 \end{bmatrix}$$

The CZ gate is a two-qubit gate which performs a conditional phase shift on the target qubit based on the state of the control qubit [43, 44]. If the control qubit is $|1\rangle$ then there will be phase shift otherwise no phase shift is applied [45, 46]. This is used for achieving entanglement.

C. Accuracy

$$\text{Accuracy} = \frac{TN + TP}{TN + FN + FP + TP} \tag{2.4}$$

Equation (2.4) represents the overall accuracy of a model which is the proportion of prediction that the model got right. It should be as high as possible.

D. Recall

$$\text{Recall} = \frac{TP}{TP + FN} \tag{2.5}$$

Equation (2.5) represents the Recall or TPR which is the proportion of positive cases that the model identified correctly. It should also be as high as possible to validate the accuracy of working model.

E. Precision

$$\text{Precision} = \frac{TP}{TP + FP} \tag{2.6}$$

Equation (2.6) shows the precision which is the proportion of predicted positive cases where the true label is actually positive. Similar to accuracy and recall it should also be high to indicate the performance of good model.

F. F-1 Score

$$\text{F-1 Score} = \frac{2{*}\text{Precision}{*}\text{Recall}}{\text{Precision} + \text{Recall}} \tag{2.7}$$

Equation (2.7) gives the F-1 score which is derived from precision and recall together. It balances the trade-off between precision and recall. Its value varies from 0 to 1. The higher the value, higher will be the performance.

2.5 Results and Discussion

2.5.1 Numerical Analysis

The numerical analysis will explore all the insights related to numbers and statistics. This will help in understanding market factors like “Relative Strength Index” and performance metrics like accuracy.

Table 2.2 Relative strength index of data

Dataset row	RSI value
1	100.00
2	43.24
3	43.24
..	
..	
..	
..	
..	
632,217	48.13
632,218	47.85
632,219	50.45

2.5.1.1 Relative Strength Index

The Relative Strength Index (RSI) is a special tool used a lot when people look at stocks. This will help us see how strong and fast the price of a stock is changing. Traders and analysts use it to understand if a stock is doing well or not. For us as well it will help uslook at the gains and losses of the stock over a certain time. It is like a way to figure out how powerful and quick the changes in the stock's price are. It is given as follows:

$$\text{RSI} = 100 - \frac{100}{1 + RS} \tag{2.8}$$

where, $RS = \frac{Average\ gain}{Average\ loss}$.

Table 2.2 presents the RSI values for the given dataset ranging from 0 to 100. An RSI crossing 70 is categorized as "Overbought", suggesting potential overpricing and anticipating a price correction or reversal. Conversely, an RSI falling below 30 is labeled as "Oversold" indicating potential underpricing and a possible upward price correction. The table illustrates a mix of overpriced and underpriced stocks with an average RSI of approximately 45.28.

2.5.1.2 Performance Metrics

Variational quantum circuit is a supervised learning used for classification of data [47, 48]. Metrics such as accuracy, recall, precision etc. are used to validate the performance of the model. The following table will tabulate all the results obtained during the training process.

Table 2.3 shows the performance metrics of the proposed model under different circumstances. In the first scenario where all the major attributes were considered had showed the flawless results. In the second case, there are some false positive but there

Table 2.3 Performance analysis

S. no.	Conditions (with COBYLA)	Performance metrics			
		Precision (in %)	Recall (in %)	F-1 Score (0–1)	Specificity (0–1)
1	4 Qubits with real amplitudes	100	100	1	1
2	2 Qubits with real amplitudes	98	97	0.98	0.97
3	2 Qubits with EfficientSU2	99	99	0.99	0.99

is a balance between precision and recall. In the final case where two qubits were considered along with EfficientSU2 optimizer, the performance is almost equivalent to the scenario where four qubits were taken. So it is pretty obvious to consider the one where the accuracy is high and the number of qubits passed is less making the model more optimal and hence superior for forecasting stock.

2.5.2 Graphical Analysis

Figure 2.7a, b, c above illustrates the objective function (or cost function) against the iteration along with the time taken to train the model under three conditions. It could be observed that the cost function for Fig. 2.7b, c and the time taken by them is optimal and hence are good to choose for prediction. Also, compared to the first model the training time deceases by almost 9 times making the quantum circuit apparatus more appealing. When we visualize the convergence, Fig. 2.7b reached the stabilized state earlier while Fig. 2.7c did not achieve the convergence at all making the second scenario-based circuit a promising model for our goal.

Figure 2.8a, b, c represents the ROC curve for different quantum circuit cases. The AUC-ROC curves are integral for evaluating the performance of classification models which provide insightful observations in the context of our thesis. The first curve demonstrating an AUC-ROC value of 1, signifies a model achieving ideal discrimination between classes and showcasing impeccable sensitivity and specificity. The second curve with an ROC value of 0.97 indicates a high level of accuracy and effectiveness in distinguishing between positive and negative instances. Notably the third curve boasts an exceptional ROC value of 0.99, underscoring the model's outstanding discriminatory power. These values collectively emphasize the robust performance of our classification models, showcasing their ability to discern between classes with high precision and reliability.

Figure 2.9 above illustrates the bar graph showing the comparative analysis of classical [49–51] and quantum model along with different sets of features and optimizers. It could be perceived that the quantum model performs equivalent to the classical one and hence shows the promise to predict stock data based on time series.

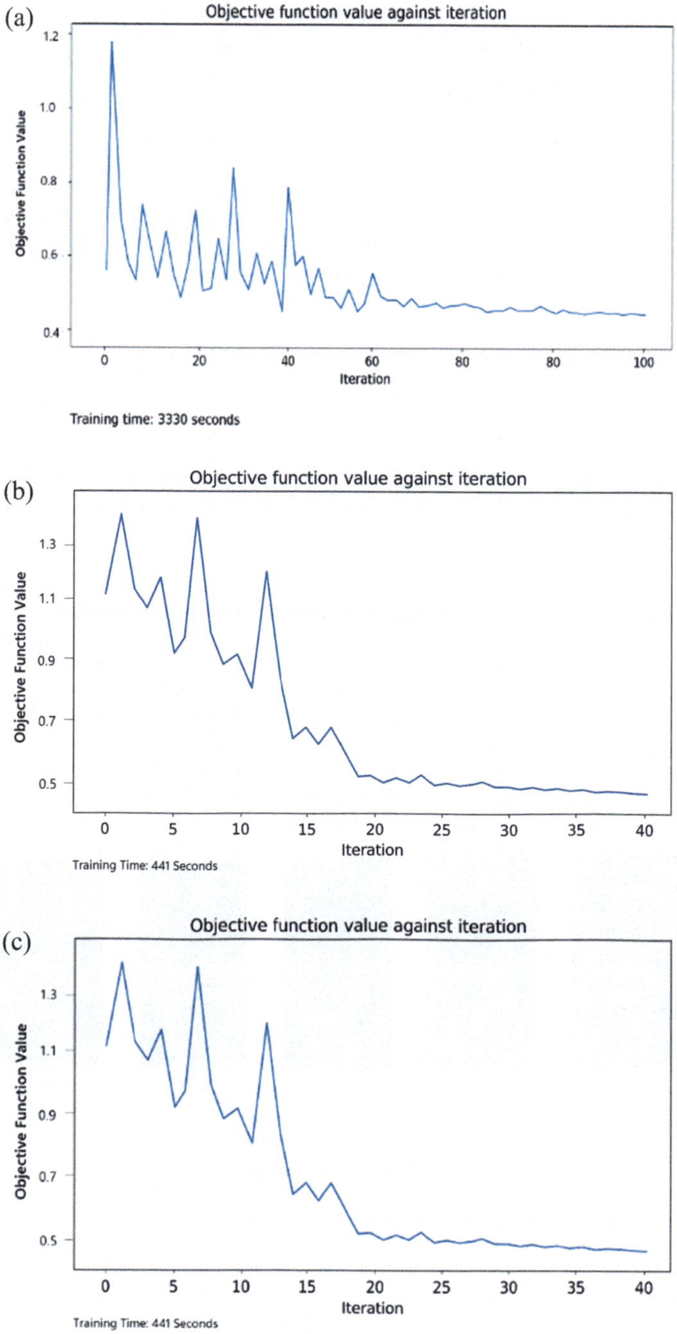

Fig. 2.7 **a** Cost function for four qubits with Real Amplitudes optimizer. **b** Cost function for two qubits with Real Amplitudes optimizer. **c** Cost function for two qubits with EfficientSU2 optimizer

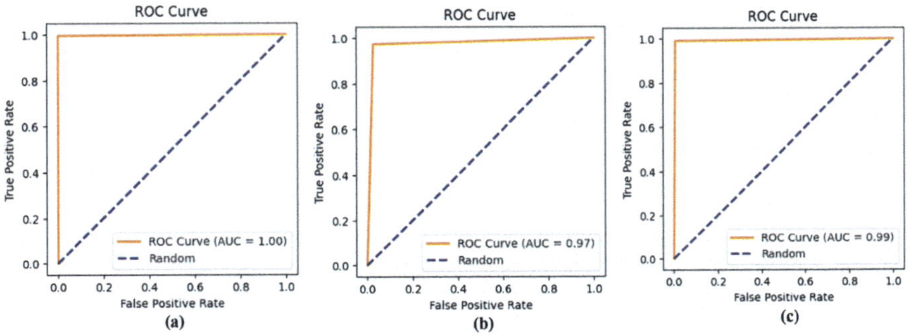

Fig. 2.8 **a**, **b**, **c** AUC-ROC Curve for different variational quantum circuit

By considering the overall analysis we could conclude that the highly correlated data could be easily predicted with just two features and the optimizers plays a vital role as well in finding optimal parameters for parameterized circuits like variational quantum circuit.

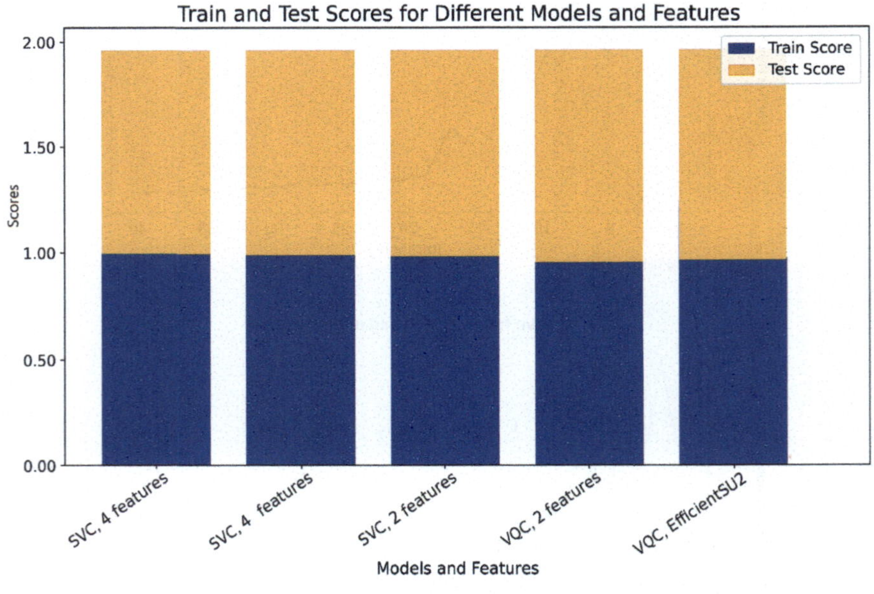

Fig. 2.9 Bar graph showing training and testing score for different apparatus

2.6 Conclusions

While the understanding of quantum computing and its algorithms is currently in a developmental phase, the potential application of quantum computing in stock prediction is a promising prospect. Notably, an almost 99% overall accuracy was achieved in cases where time series data exhibited high correlation. Additionally, the Real Amplitudes optimizer proves to be effective in discovering optimal parameters and reducing processing time by minimizing the cost function. Managing the escalating volume of data poses a significant challenge and this is precisely where quantum machine learning emerges as a crucial solution and capable of swiftly processing vast amounts of data. Quantum learning, although when dealing solely with numerical data becomes even more complex when extended to linguistic elements such as sentiments. As major tech giants actively pursue advancements in quantum platforms it is anticipated that user-friendly approaches for handling quantum data will be developed. Despite existing limitations in stock market prediction and deprecated libraries there is a steadfast commitment to exploring this intriguing domain that poised to unveil breakthroughs. The integration of quantum machine learning in stock prediction is anticipated to progress as quantum computing technology becomes more accessible although mainstream adoption in the finance industry may require some time to materialize.

References

1. Nielsen, M.A., Chuang, I.L.: Quantum Computation and Quantum Information. Cambridge University Press (2010)
2. Rieffel, E., Polak, W.H.: Quantum Computing: A Gentle Introduction. MIT Press (2011)
3. Wittek, P.: Quantum Machine Learning: What Quantum Computing means to Data Mining. Academic (2014). ISBN: 9780128009536
4. Yanofsky, N.S., Mannucci, M.A.: Quantum Computing for Computer Scientist. Cambridge University Press (2008)
5. Kopczyk, D.: Quantum machine learning for data scientists (2018). https://doi.org/10.48550/arXiv.1804.10068
6. Abhijith, J., et al.: Quantum algorithm implementations for beginners. Assoc. Comput. Mach. (ACM) **3**(4), 1–92 (2022)
7. Saxena, A., Mancillia, J., Montalban, I., Pere, C.: Finacial Modelling using Quantum Computing, Packt (2023). ISBN:9781804618424
8. Divya, R., Peter, J.D.: Quantum Machine Learning: A Comprehensive review on Optimization of Machine Learning Algorithms, pp. 1–6. IEEE (2021)
9. Cerezo, M., Verdon, G., Huang, H.Y., Cincio, L., Coles, P.J.: Challenges and opportunities in quantum machine learning. Nat. Comput. Scie. **2**(567–576) (2022). https://doi.org/10.1038/s43588-022-00311-3
10. Garcia, D.P., Benito, J.C., Penalvo, F.J.: Systematic literature review: quantum machine learning and its applications (2023). https://doi.org/10.48550/arXiv.2201.04093
11. Yamasaki, H., Isogai, N., Murao, M.: Advantage of Quantum Machine Learning from General Computational Advantages (2023). https://doi.org/10.48550/arXiv.2312.03057

12. Kobayashi, N., Suimon, Y., Miyamoto, K., Mitarai, K.: The Cross-sectional Stock Return Predictions via Quantum Neural Network and Tensor Network (2023). https://doi.org/10.48550/arXiv.2304.12501
13. Mahajan, R.P.: Stock price prediction using quantum neural network. J. Global Res. Comput. Sci. **1**(4) (2010)
14. Cao, Y., Zhou, X., Fei, X., Zhao, H., Liu, W., Zhao, J.: Linear-layer-enhanced quantum long short-term memory for carbon price forecasting (2023). https://doi.org/10.1007/s42484-023-00115-2
15. Srivastava, N., Gaurang, B., Neel, S., Aswath, H.: The Potential of Quantum Techniques for Stock Price Prediction (2023). https://doi.org/10.48550/arXiv.2308.13642
16. Alaminos, D., Salas, M.B., Gámez, M.A.: Forecasting stock market crashes via real-time recession probabilities: a quantum computing approach. World Sci. (2022). https://doi.org/10.1142/S0218348X22401624
17. Paquet, E., Soleymani, F.: QuantumLeap: Hybrid quantum neural network for financial predictions (2022). https://doi.org/10.1016/j.eswa.2022.116583
18. Liu, G., Ma, W.: A quantum artificial neural network for stock closing price prediction. LSEVIER (20220. https://doi.org/10.1016/j.ins.2022.03.064
19. Dimoska, S., Emmanoulopoulos, D.: Quantum Machine Learning in Finance: Time Series Forecasting (2022) https://doi.org/10.48550/arXiv.2202.00599
20. Marco, P., Ahmad, S.F., Ajagekar, A., Buts, A., Chakrabarti, S., Herman, D., Hu, S.: Quantum Machine Learning for Finance (2021). https://doi.org/10.48550/arXiv.2109.04298
21. Alaminos, D., Salas, M.B., Gámez, M.A.: Quantum Computing and Deep Learning Methods for GDP Growth Forecasting (2022). https://doi.org/10.1007/s10614-021-10110-z
22. Adharsh, B., Adithya, P., George, A., Krrish, K., Mridul, S., Showmen, T., Larry, M.: Near Future Stock Market Forecasting Based on Chaos Theory, Sentiment Analysis, and Quantum Computing. ASDRP (2020)
23. Ahmed, A.S., Kurnaz, S.: Quantum computing and artificial neural network for classification and time series prediction. In: International Congress on Human-Computer Interaction, Optimization and Robotic Applications (HORA), pp. 1–4 (2022)
24. Guo, M., Weng, Y., Ye, L., Lai, Y.C.: Continuous variational quantum algorithms for time series. In: International Joint Conference on Neural Networks (IJCNN), pp. 1–8 (2020)
25. Chimprang, N., Tansuchat, R.: An application of quantum optimization with fuzzy inference system for stock index futures forecasting. In: International Conference of the Thail and Econometrics Society (2020). https://doi.org/10.1007/978-3-030-97273-8_27
26. Wang, C., Yang, Y., Xu, L., Wong, A.: A Hybrid Model of Primary Ensemble Empirical Mode Decomposition and Quantum Neural Network in Financial Time Series Prediction. World Scientific Publishing Company (2020). https://doi.org/10.1142/S0219477523400060
27. Huggins, W., Patil, P., Mitchell, B., Whaley, K.B., Stoudenmire, E.M.: Towards quantum machine learning with tensor flow. Quantum Sci. Technol. (2019). ISSN:2058-9565
28. Takaki, Y., Mitarai, K., Negoro, M., Fujii, K., Kitagawa, M.: Learning temporal data with variational quantum recurrent neural network (2020). https://doi.org/10.48550/arXiv.2012.11242
29. Killoran, N., Brombley, T.R., Arrazola, J.M., Schuld, M., Quesada, N., Lloyd, S.: Continuous-variable quantum neural networks (2018). https://doi.org/10.48550/arXiv.1806.06871
30. Yasunari, S., Kawase, Y., Masumura, Y., Hiraga, Y., Nakadai, M., Chen, J., Nakanishi, K.M.: Qulacs: A Fast and Versatile Quantum Circuit Simulator for Research Purpose (2020). https://doi.org/10.22331/q-2021-10-06-559
31. Carlos, A., Tiago, F.: Time series forecasting with qubit neural networks. In: International Conference on Artificial Intelligence and Soft Computing (ICAISC) (2017)
32. Ville, B., Izaac, J., Schuld, M., Gogolin, C., Ahmed, S., Ajith, V., Alam, M.S.: PennyLane: Automatic Differentiation of Hybrid Quantum-classical Computations (2018)
33. Wang, G.: Quantum algorithm for linear regression (2017). https://doi.org/10.1103/PhysRevA.96.012335

34. Ruiz, R., Vazquez, M., Romero, L.: Time series forecasting with quantum machine learning architectures. In: Mexican International Conference on Artificial Intelligence, pp. 66–82 (2022)
35. Kariya, A., Behera, B.K.: Investigation of quantum support vector machine for classification in NISQ era (2021). https://doi.org/10.48550/arXiv.2112.06912
36. Weinstein, M., Horn, D.: Dynamic quantum clustering: a method for visual exploration of structures in data. PubMed (2009). https://doi.org/10.1103/PhysRevE.80.066117
37. Khan, T.M., Kelly, A.R.: Machine learning: quantum versus classical. **8**(3), 219275–219294 (2020). IEEE
38. Manjunath, C., Marimuthu, B., Ghosh, B.: Analysis of nifty 50 index stock market trends using hybrid machine learning model in quantum finance. IJECE **13**(3), 3549–3560 (2023)
39. Sosa, D.S., Telahun, T., Elmaghraby, A.: TensorFlow quantum: impacts of quantum state preparation on quantum machine learning performance. **8**(4), 215246–215255 (2020). IEEE
40. Chen, S., Yang, C., Qi, J., Chen, P., Ma, X., Goan, H.: Variational quantum circuits for deep reinforcement learning. **8**(3), 141007–141024 (2020). IEEE
41. Liu, W., Gao, P., Wang, Y., Yu, W., Zhang, M.: A unitary weights based one-iteration quantum perceptron algorithm for non-ideal training sets. **7**(1), 36854–36865 (2019). IEEE
42. Biamonte, J., Wittek, P., Pancotti, N., Rebentrost, P., Wiebe, N., Lloyd, S.: Quantum machine learning. Nature **549**(15), 195–202 (2017)
43. Aimeur, E., Brassard, G., Gamps, S.: Quantum speed-up for unsupervised learning. **90**(2), 261–287 (2013). Springer
44. McClean, J.R., Boixo, S., Smelyanskiy, V.N., Babbush, R., Neven, H.: Barren plateaus in quantum neural network training landscapes. Nat. Commun. **9**(1), 1–6 (2018)
45. McClean, J.R., Romero, J., Babbush, R., Guzik, A.: The theory of variational hybrid quantum-classical algorithms. New J. Phys. **18**(2) (2016)
46. Egger, D.J., et al.: Quantum computing for finance: state-of-the-art and future prospects. IEEE Trans. Quantum Eng. **1**(5), 1–24 (2020)
47. Rebentrost, P., Gupt, B., Bromley, T.R.: Quantum computational finance: Monte Carlo pricing of financial derivatives. Am. Phys. Rev. **98**(2) (2018)
48. Alcazar, J., Ortega, V.L., Ortiz, A.P.: Classical versus quantum models in machine learning: insights from a finance application. **1**(3) (2020). IOP Publishing
49. Ryan, C.: Quantum Computing and Quantum Machine Learning: Harnessing the Power of Quantum Mechanics for Intelligent Data Science. Cambridge University Press (2012)
50. Hidary, J.D.: Quantum Computing: An Applied Approach. Springer (2021)
51. Jadhav, A., Rasool, A., Gyanchandani, M.: Quantum machine learning: scope for real-world problems. Int. Conf. Mach. Learn. Data Eng. **218**(2612–2625) (2023). https://doi.org/10.1016/j.procs.2023.01.235

Chapter 3
Quantum Dot Cellular Automata: Breaking Barriers in Electronics Circuitry for Tomorrow's Technologies

Supreeti Kamilya and Soumyadeep Paty

Abstract The chapter explores the revolutionary potential of Quantum Dot Cellular Automata (QCA) as a groundbreaking technology in electronic circuitry, which could succeed traditional CMOS systems. QCA stands out for its significant advantages, offering greater efficiency, higher operational speed, and lower power consumption, making it a compelling candidate for next-generation electronic circuitry. The chapter outlines the evolution of QCA technology, beginning with basic gate designs and progressing to more complex components such as adders, flip-flops, and multiplexers. This progression highlights QCA's potential as a viable alternative to current semiconductor technologies. QCA operates on the principles of quantum mechanics, utilizing quantum dots as the basic elements for information processing. This innovative approach allows for ultra-low power consumption and high-speed operations, with scalability that surpasses conventional silicon-based technologies. The chapter explores the practical applications of QCA in different digital logic circuits which can be used in computing systems. While QCA is still in the research phase and has yet to see widespread implementation, its potential to revolutionize electronic circuitry is clear. This chapter provides an outline of QCA's transformative capabilities, positioning it as a key technology for future advancements in electronics.

Keywords QCA · Ultra-low power electronics · New generation electronics circuitry · Logic gates · Transformative technologies · Quantum computing

S. Kamilya
Department of Computer Science and Engineering, Birla Institute of Technology, Mesra, Ranchi, India
e-mail: supreeti.kamilya17@bitmesra.ac.in

S. Paty
Department of Mining Engineering, Kazi Nazrul University, Asansol, India

S. Paty (✉)
Centre for IoT and AI Integration with Education-Industry-Agriculture, Kazi Nazrul University, Asansol, India
e-mail: mining.soumyadeep@knu.ac.in

S. Dehuri et al. (eds.), *Advances in Quantum Inspired Artificial Intelligence*, Intelligent Systems Reference Library 274, https://doi.org/10.1007/978-3-031-89905-8_3

3.1 Introduction

Quantum Dot Cellular Automata (QCA) represents a significant departure from traditional CMOS technology, offering a novel approach to information processing, storage, and manipulation in electronic circuits. As the limitations of CMOS technology become more apparent, particularly in terms of power consumption and miniaturization, QCA emerges as a potential successor capable of addressing these challenges [1]. Traditional electronic circuits have long relied on the use of transistors, which operate by switching between on and off states to represent binary data, 0 and 1 s. This approach, while highly successful, is increasingly constrained by the physical limitations of silicon-based materials, particularly as the industry pushes towards smaller and more power efficient devices. However, as predicted by Moore's Law—the observation that the number of transistors on a chip doubles approximately every 18 months—the semiconductor industry is approaching the physical and technical limits of CMOS technology [2]. Whereas QCA is a nano electronic technology that uses quantum–mechanical effects to encode and process information. Unlike conventional circuits that rely on electrical currents, QCA manipulates electron positioning within quantum dots to represent binary data. The operation of QCA circuits is based on electron tunnelling and Coulombic interactions. When a QCA cell's polarization changes, it influences the polarization of its neighbouring cells, allowing binary signals to propagate without requiring electrical current flow. This cascading effect enables efficient and high-speed data transmission. The primary components of a QCA circuit include QCA wires, which propagate binary signals, inverters (NOT gates) that flip cell polarization, majority gates that perform logical operations, and a four-phase clocking mechanism that controls data flow and synchronization. One of the advantages of QCA is its ultra-low power consumption. Unlike CMOS transistors, which consume power due to leakage currents and switching activities, QCA circuits rely on the static positioning of electrons, significantly reducing energy dissipation. Additionally, QCA-based designs have the potential to operate at sub-threshold voltage levels, making them ideal for energy-efficient computing applications. Another key advantage of QCA is its high-speed operation. The absence of electrical current flow eliminates resistance and capacitance effects, allowing information to propagate through electrostatic interactions almost instantaneously. This leads to faster computation times compared to traditional semiconductor-based circuits. Scalability is also a major strength of QCA. The small size of QCA cells allows for highly compact and dense circuit designs, offering a viable alternative as CMOS technology struggles to scale below 5 nm. As CMOS nears saturation, the search for new and more advanced technologies has become critical. QCA offers a radical departure from this technology by utilizing the principles of quantum mechanics to achieve similar, and often superior, results without the need for traditional transistors [3–6]. Quantum dots are tiny semiconductor structures designed to trap and control electrons. In QCA, binary information is encoded by arranging electrons within a grid of quantum dots instead of moving charges. This innovative method enables QCA circuits to operate with ultra-low power consumption and high-speed performance, making it a compelling candidate for the next generation of electronic devices [7].

The potential applications of QCA span various domains of digital electronics. QCA can be used to design arithmetic circuits such as adders and multipliers, which are more efficient and compact than their CMOS counterparts. Additionally, QCA-based memory units offer high-speed access with minimal power requirements, making them ideal for next-generation computing systems. Multiplexers and decoders, which are crucial in data routing and signal processing, can also be efficiently implemented using QCA technology. Moreover, sequential circuits, including flip-flops and counters, can be designed using QCA to enhance data storage and state transitions in computing systems. Furthermore, as a stepping stone toward quantum computing, QCA has the potential to enable future advancements in nanocomputing and high-performance processors.

This chapter begins by introducing the basic principles of QCA, including the structure and operation of its fundamental unit, the QCA cell. It then explores the implementation of basic logic gates, AND, OR & NOT, using QCA, highlighting the efficiency and compactness of QCA-based designs. The chapter further delves into more complex circuit designs, including adders, multiplexers, and sequential circuits like flip-flops, showcasing QCA's potential to revolutionize electronic circuitry. Through these explorations, the chapter underscores QCA's capacity to overcome the inherent limitations of current semiconductor technologies, paving the way for future advancements in electronics.

3.2 QCA Technology

QCA is a groundbreaking innovation poised to revolutionize electronic circuitry by fundamentally changing how information is processed, stored, and manipulated. Central to QCA are quantum dots, nanoscale semiconductor structures capable of confining electrons. In a QCA cell, the position of electrons within these quantum dots determines the binary state of the system. Unlike conventional circuits that move charges through wires, QCA encodes information based on the spatial arrangement of electrons within a grid of quantum dots. This approach allows QCA circuits to operate with ultra-low power consumption, a critical advantage as energy efficiency becomes increasingly important in electronics design.

The four component of QCA is the QCA cell (qubit). It comprises 4 quantum dots arranged in a square configuration. Here, 2 electrons occupy 2 dots among the 4 dots. The two electrons will naturally repel each other due to Coulombic repulsion and will settle in a configuration where they are as far apart as possible. The two stable configurations of the electrons are diagonal, leading to two possible polarization states: One state represents a binary '0' (when the electrons are in one diagonal pair of dots) (Fig. 3.1). The other state represents a binary '1' (when the electrons are in the opposite diagonal pair) (Fig. 3.1) [8]. This diagonal placement is crucial for encoding binary information in QCA. The polarization of each cell influences neighbouring

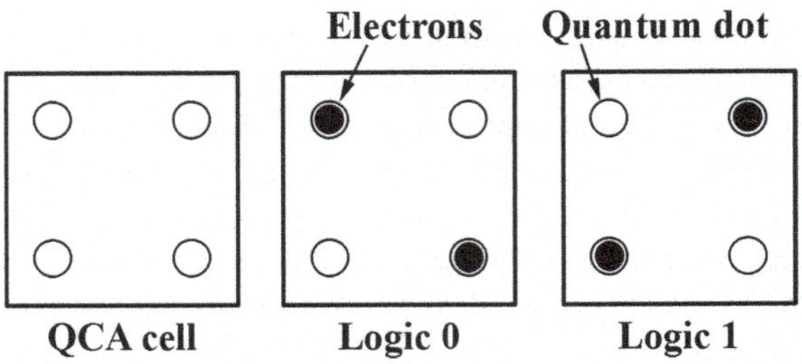

Fig. 3.1 QCA cell with different polarizations

cells, allowing the transfer of information through the QCA network. The interaction between adjacent cells in a QCA array happens because the polarization of one cell affects the polarization of its neighbours, causing them to align. The diagonal placement of electrons maximizes this interaction.

3.3 Basic Gate Implementation Using QCA

In QCA, a wire is composed of a series of aligned QCA cells that transmit binary information by propagating polarization states. Each cell in the wire influences its neighbours, allowing a binary signal, imposed by the input cell to cascade through the wire. Unlike traditional wires that rely on current, QCA wires transmit signals through quantum tunnelling and electrostatic interactions between cells, enabling efficient and low-power signal propagation across the circuit. These wires are fundamental in linking various components within QCA circuits, such as gates and logic units, to enable complex computational processes. Figure 3.2 represents the wire where the QCA cells are cascaded and transmit binary information [9].

QCA can be used to design digital logic circuits that are more efficient and compact than their conventional semiconductor-based circuits. Conventional digital circuits utilize transistors to execute logic operations by regulating current flow. In contrast, QCA encodes and processes binary information by manipulating electron positions within quantum dots. In any QCA circuit design, a cell is placed at the input port to govern the behaviour of wires and logic gates.

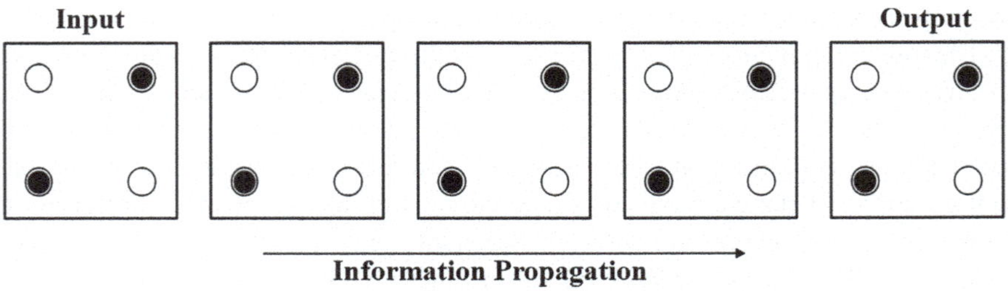

Fig. 3.2 QCA wire

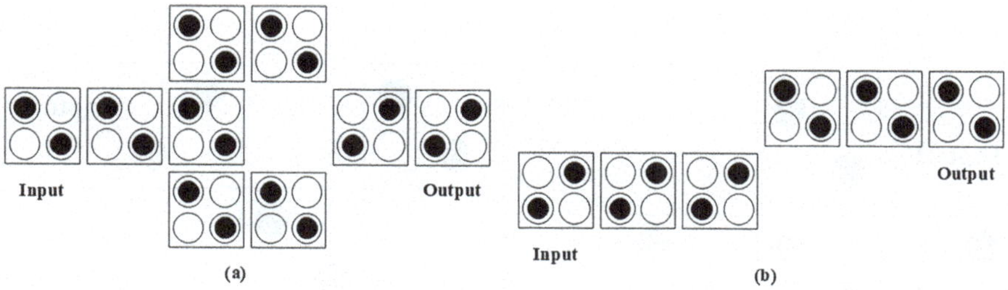

Fig. 3.3 **a** Traditional QCA robust inverter **b** QCA simple inverter

AND, OR, and NOT, are the basic gates in digital logic whereas NAND and NOR are the universal gates. These gates are the foundational building blocks for designing complex digital circuits. QCA implements these gates using arrangements of QCA cells, where the binary states (0/1) are encoded by the electrons' positions within the cells. The foundational building blocks of QCA include three essential elements: *Wire, Inverter, and Majority Gate (MG).* These elements serve as the core components for constructing different logic functions. By combining Inverters and Majority Gates under specific logical conditions, logic gates, such as OR and AND gates, can be created, which are essential for performing various computational tasks within QCA circuits.

Inverter (NOT Gate): In a QCA inverter, the input cell's polarization determines the output cell's polarization in a manner that produces the opposite binary state. Specifically, the configuration of the quantum dots causes an inversion of the binary logic; if the input represents a binary '1', the output will be a binary '0', and vice versa. This inversion occurs due to the electrostatic interactions between neighbouring cells, where the alignment of electrons in the quantum dots forces the adjacent cell to adopt the opposite polarization [10]. Two widely used inverters implemented by QCA are shown in Fig. 3.3.

QCA inverters perform the essential logic function of signal inversion with high speed and minimal power consumption, making them a key building block in QCA-based digital circuits. Unlike traditional electronic circuits where an inverter changes the voltage level to invert the signal, a QCA inverter operates based on the arrangement of quantum dots within a cell.

Majority Gate: A Majority Gate in QCA comprise 5 closely positioned quantum cells, where 3 of the cells serve as inputs and the remaining two as intermediaries and the output. The output reflects the state of the Majority of the input cells—if two or more of input cells are polarized in one state (either '0' or '1'), the output cell will adopt that state. The Majority gate and its corresponding truth table is shown in Fig. 3.4 and Table 3.1 respectively [11].

Using the Majority gate, one can design various logic functions, including AND, OR, NAND, and NOR. Each of the gates are realized as follows.

By setting one of the Majority Gate's inputs to '0', the gate will perform an AND operation. The output will be '1' only when both of the remaining inputs are '1'.

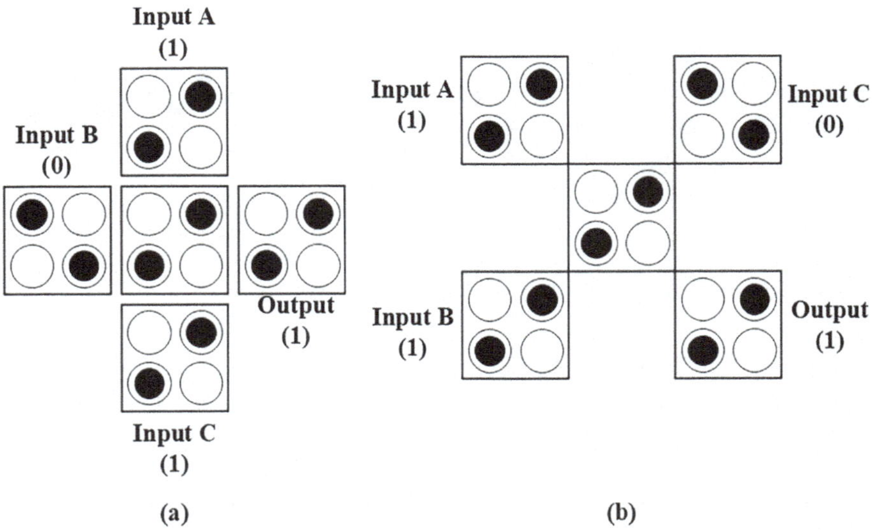

Fig. 3.4 **a** Majority gate with three input, **b** Rotated majority gate with three input

Table 3.1 Majority gate truth table

Input			Output
A	B	C	M
0	0	0	0
0	0	1	0
0	1	0	0
0	1	1	1
1	0	0	0
1	0	1	1
1	1	0	1
1	1	1	1

On the other hand, by setting one of the Majority Gate's inputs to '1', the gate will perform an OR operation. The output will be '0' only when both of the other inputs are '0'. The implementation of AND and OR gates are shown in Fig. 3.5.

To implement a NAND gate, the Majority Gate is configured similarly to the AND gate, but with an additional inversion of the output. The NAND operation produces an output of '0' only when all inputs are '1', and '1' otherwise. For a NOR gate, the Majority Gate is set up similarly to the OR gate, but with an inversion of the output. The NOR operation results in an output of '1' only when all inputs are '0', and '0' otherwise (Fig. 3.6).

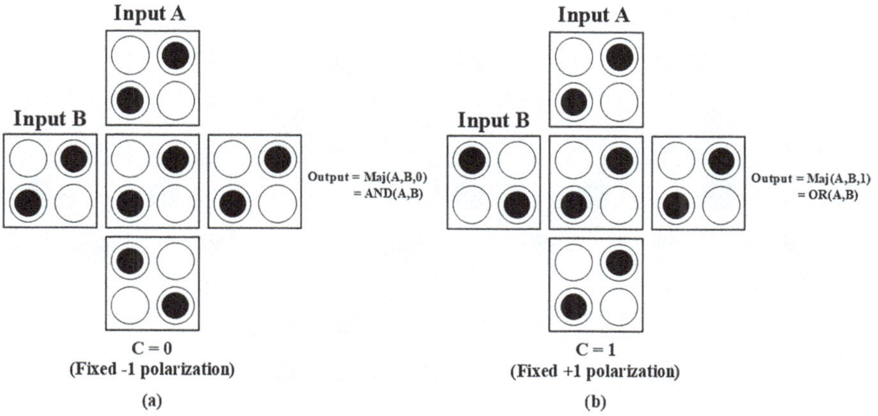

Fig. 3.5 **a** AND gate using QCA, **b** OR gate using QCA

3.4 Implementation of Combinational Circuits Using QCA

Combinational logic circuits are circuits where the output is determined solely by the current inputs, without any memory or feedback loops. QCA is well-suited for designing circuits such as adders and multiplexers as it is easy to implement the basic gates and universal gates using QCA. Half adder, the most basic adder, adds 2 single-bit binary numbers and produces a sum and a carry output. For a full-adder, which adds three single-bit numbers (two significant bits and a carry bit from a previous addition) and results a sum and a carry output, the design is slightly more complex.

Half-Adder Implementation:

The sum output of a half-adder can be implemented using an XOR gate, which can be constructed using a combination of Majority gates. By configuring MG with specific inputs and utilizing inverters to achieve the XOR functionality (Fig. 3.7), the sum is generated based on the inputs. On the other hand, the carry output is achieved using an AND gate. This can be implemented directly with a MG by setting one of its inputs to '0' and arranging the other two inputs to perform the AND operation. Figure 3.8 shows the half adder implemented using QCA Majority gate and inverter gate.

Full-Adder Implementation:

In case of a full adder, the sum output can be derived using two XOR gates. These XOR gate constructed using QCA based Majority gate and inverter as shown in Fig. 3.7. The first XOR gate calculates the sum of the two input bits, and the second XOR gate computes the final sum by incorporating the carry-in bit. The carry output of a full-adder involves calculating two separate carries: one from the addition of the input bits and one from the carry-in bit. This is achieved using a combination of MG configured to perform the OR operation on the carries from the two stages.

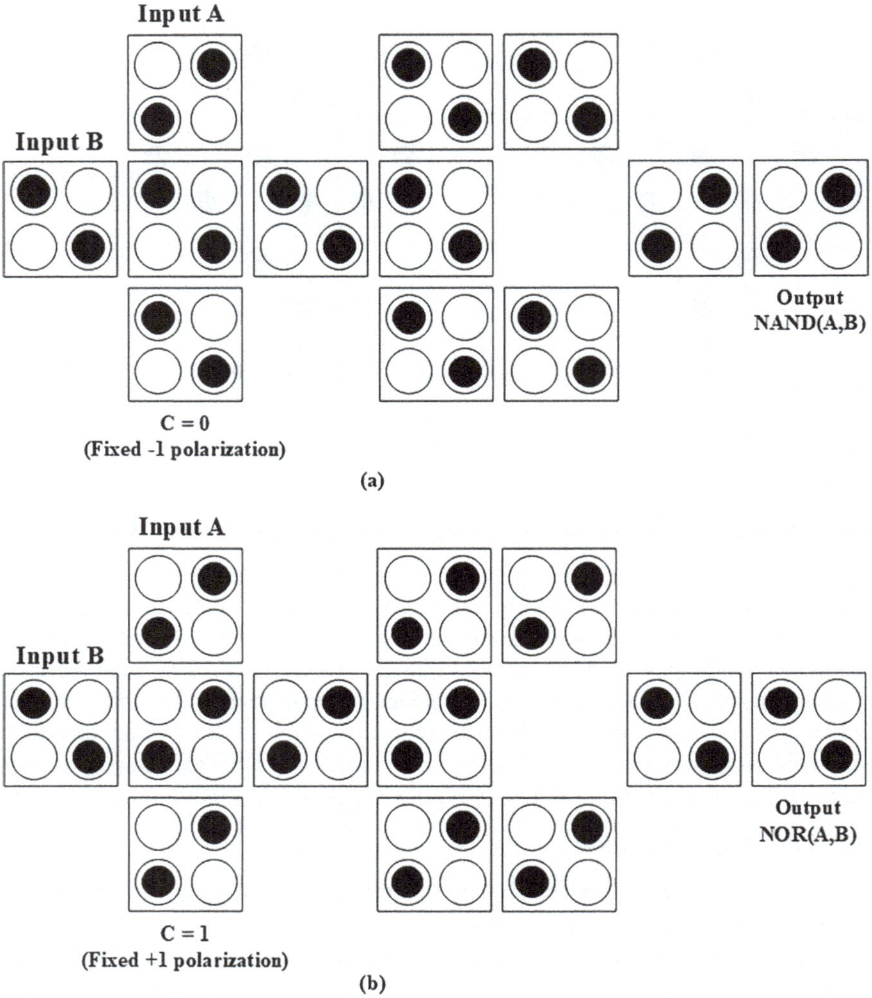

Fig. 3.6 **a** NAND gate using QCA, **b** NOR gate using QCA

Specifically, the first carry is obtained from the AND operation of the two input bits, and the second carry is derived from the OR operation of the carry-out of the first XOR gate and the carry from the second XOR gate. Figure 3.9 shows the full adder implement using QCA Majority gate and inverter gate.

Multiplexer:

For a basic 2-to-1 multiplexer, which selects one of two data inputs based on a control signal, MGs are configured to perform AND and OR operations. Specifically, the multiplexer uses MGs to calculate the contributions of each data input by combining them with the control signal and its negation. Inverters are employed to produce the

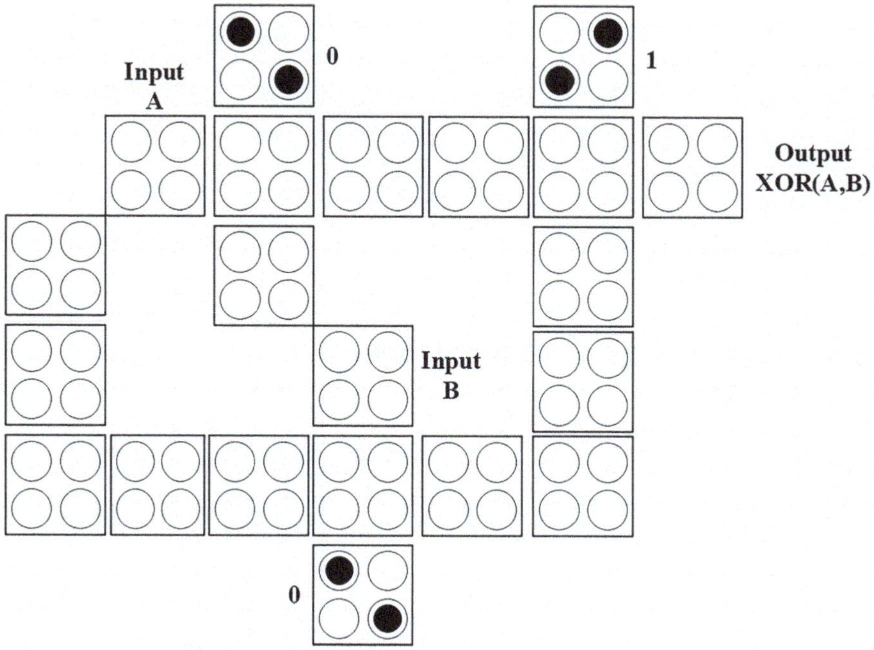

Fig. 3.7 XOR gate layout

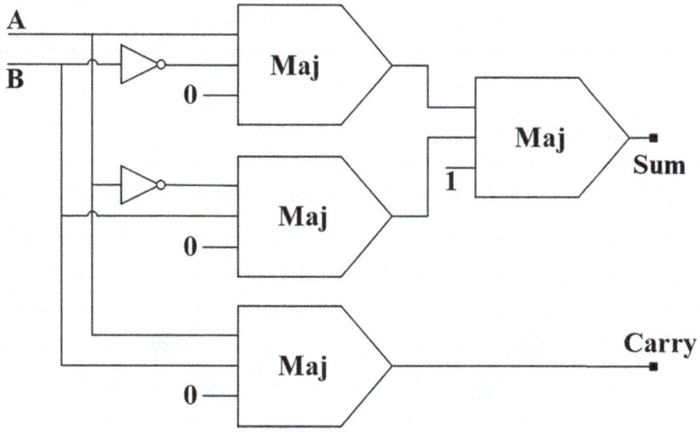

Fig. 3.8 Half adder using QCA majority gate and inverter

negation of the control signal, facilitating the correct selection between inputs [12]. The results from these AND operations are then combined using an OR-configured Majority Gate to produce the final output (shown in Fig. 3.10).

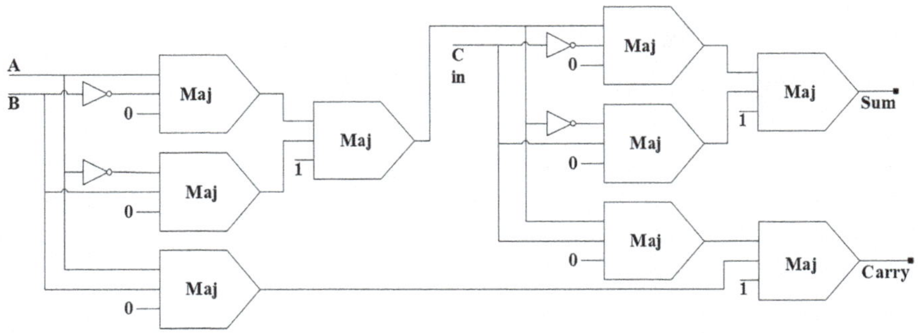

Fig. 3.9 Full adder using QCA majority gate and inverter

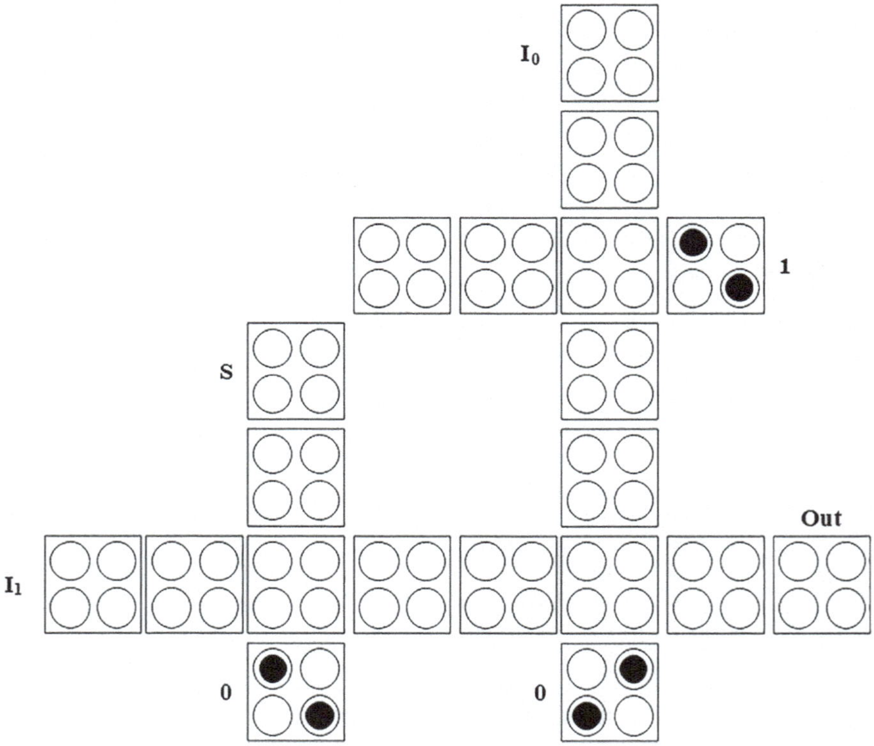

Fig. 3.10 2-to-1 multiplexer using QCA

For more complex multiplexers, such as 4-to-1 or 8-to-1, the design scales by integrating multiple levels of smaller multiplexers and additional MGs to manage a larger number of inputs and control lines.

3.5 Implementation of Sequential Circuits Using QCA

Flip-Flop:

Implementing a flip-flop with QCA involves utilizing Majority Gates and Inverters to achieve stable, bi-stable storage of binary information. The S-R flip-flop, or Set-Reset flip-flop, is one of the simplest types of digital logic circuits. It has two inputs: "Set" (S) and "Reset" (R), which control the output state. In contrast, a D flip-flop has just one data input. The D flip-flop is created by modifying the S-R flip-flop—this involves connecting the Reset (R) input through an inverter and linking the Set (S) input directly to the data input. When this adjustment is made, the result is a clocked version of the S-R flip-flop, called the D flip-flop. To construct a QCA-based D flip-flop, the design incorporates two interconnected Majority Gates arranged to form a feedback loop. One Majority Gate is used to set the state based on the data input (D) and the clock signal, while the second Majority Gate maintains the previous state, ensuring stability and data retention. Inverters are employed to handle the control signals and create the necessary feedback conditions for correct operation. The output of these Majority Gates is then used to maintain the flip-flop's state until updated by a new clock pulse. This configuration allows the QCA flip-flop to store and transition between binary states effectively. Figure 3.11 shows the implementation of S-R flip-flop and Fig. 3.12 shows a D flip-flop using QCA.

A novel QCA based S-R flip-flop has been developed by [13] where S-R flip-flop is developed using only one Majority gate. The schematic as well as the QCA layout of this S-R flip-flop is shown in Fig. 3.13. Also, a simplified D flip-flop using 3 Majority gate is proposed in [13], which is shown in Fig. 3.14.

Counters:

Implementing a counter in QCA involves constructing a series of flip-flops (typically D flip-flops) [14] that are interconnected to form a sequential circuit capable of counting binary numbers. The counter design relies on the clocked operation of flip-flops, where each flip-flop represents one bit of the counter's output. In a QCA-based counter, the first flip-flop (the least significant bit) toggles its state with every clock pulse. The subsequent flip-flops toggle their states based on the previous flip-flop's output, effectively counting in binary as clock pulses are received. Majority Gates are used within each flip-flop to manage the input and control signals, ensuring the correct toggling behaviour. Inverters help in creating the necessary feedback loops that define the state transitions. As the clock signal progresses, the output of the series of flip-flops generates a binary count, with each flip-flop contributing to a different bit of the overall count. Schamatic diagram of a 4-bit asynchronous up counter using Majority gate is shown in Fig. 3.15.

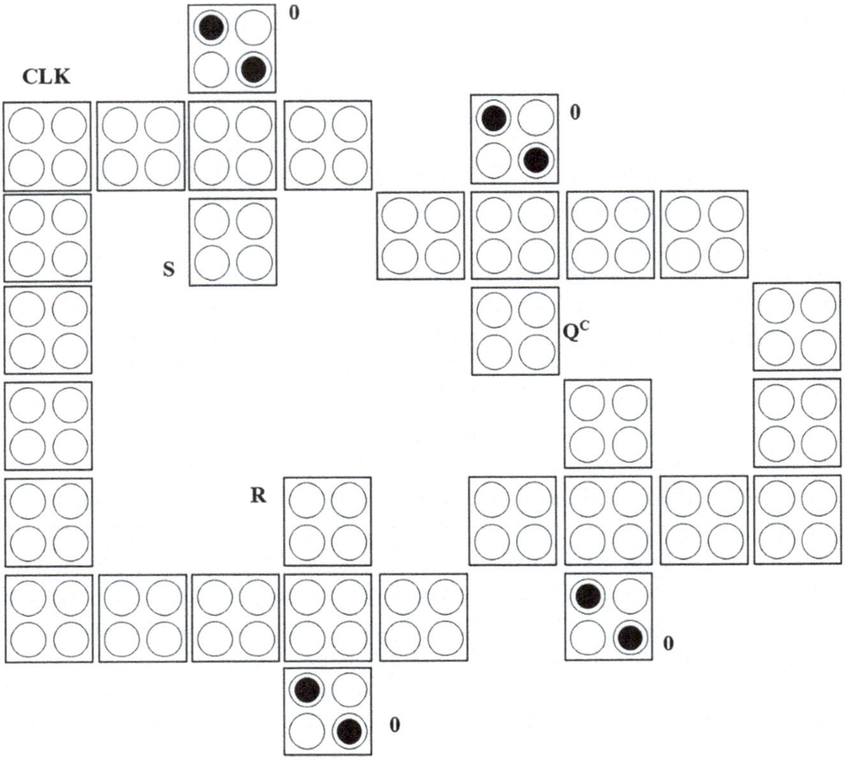

Fig. 3.11 S-R flip-flop using QCA

3.6 Conclusion

The exploration of QCA in this chapter demonstrates its revolutionary potential as an alternative to traditional CMOS technology. A close look at QCA's core principles and its use in designing both simple and advanced digital circuits highlights its key benefits, including high power efficiency, fast operation, and scalability. The unique way in which QCA manipulates binary information—through the positional arrangement of electrons in quantum dots—marks a significant shift from conventional methods, promising a new era of innovation in electronic circuit design. The gates and circuits presented in this chapter are designed with simplicity using QCA, and future enhancements could further optimize space and power consumption.

As CMOS technology nears its limitations in terms of area efficiency, exploring new technological avenues becomes imperative. QCA presents a compelling alternative, offering potential benefits in circuit design and implementation. While QCA is still in the research phase and faces challenges in widespread implementation, its demonstrated capabilities in this chapter position it as a key technology for the future of electronics. The successful design of basic gates, adders, multiplexers, and

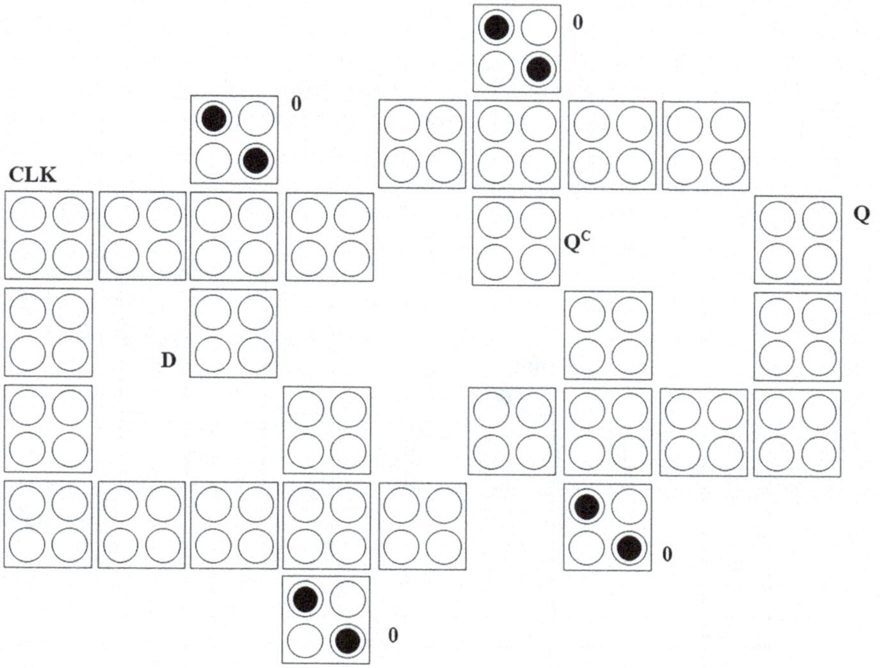

Fig. 3.12 D flip-flop using QCA

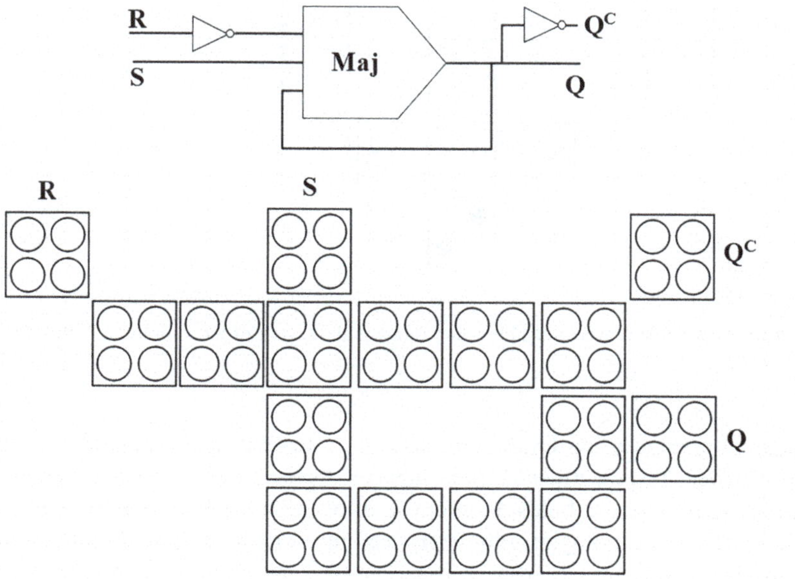

Fig. 3.13 Schematic diagram and QCA layout of the S-R flip-flop using a single majority gate

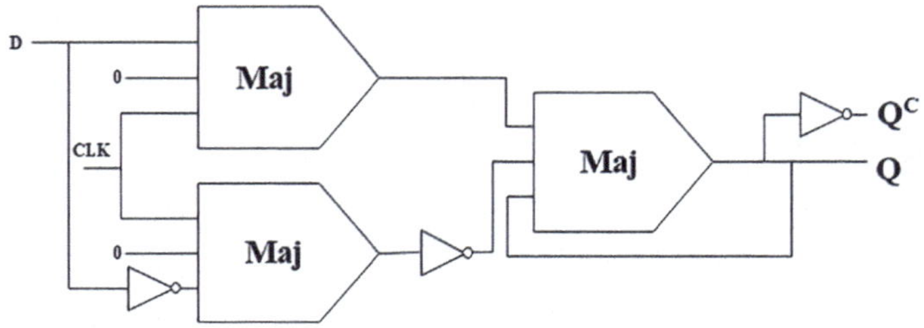

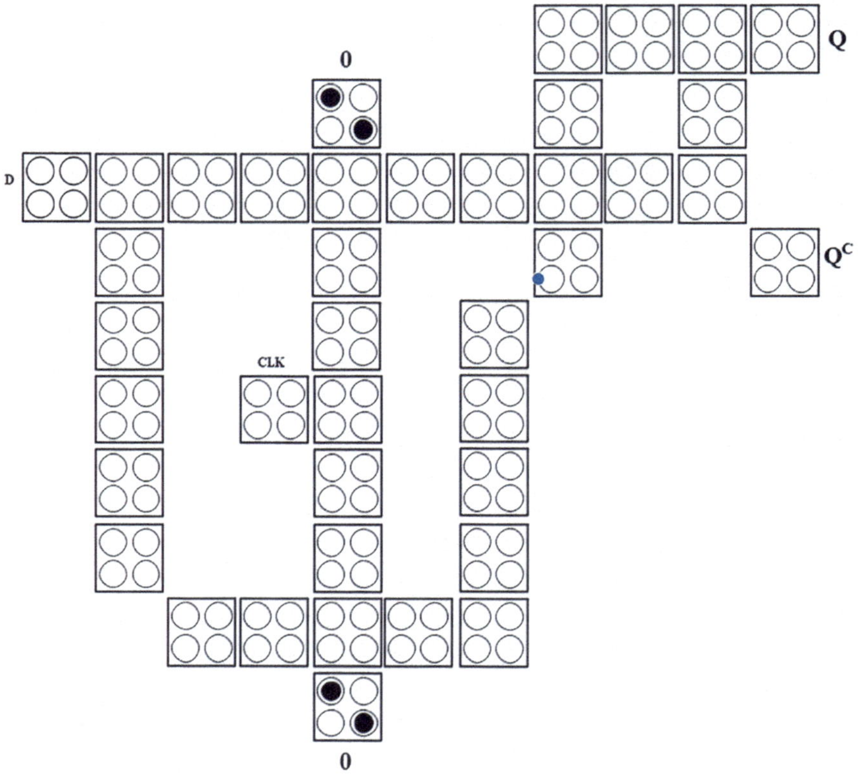

Fig. 3.14 Schematic diagram of D flip-flop (above) and layout of the simplified D flip-flop using QCA (below)

sequential circuits using QCA showcases its versatility and potential for creating more efficient and powerful electronic devices. As research and development in QCA continue to progress, it holds the promise of overcoming the limitations of current semiconductor technologies, ultimately leading to more advanced and sustainable electronic systems.

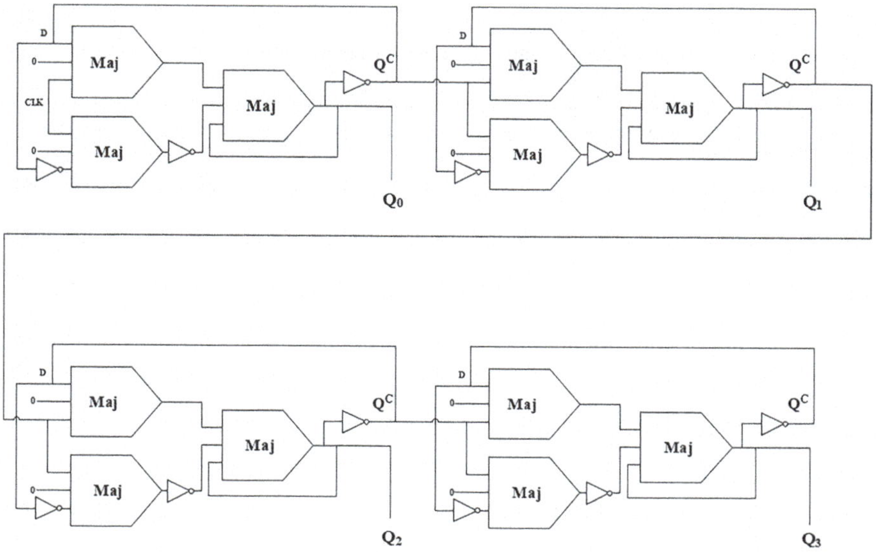

Fig. 3.15 A 4-bit asynchronous up counter using majority gate

References

1. Chetti, S.C., Yatgal, O.: QCA: a survey and design of logic circuits. Global Transitions Proc. **3**, 142–148 (2022)
2. Lent, C.S., Snider, G., Bernstein, G., Porod, W., Orlov, A., Lieberman, M., Fehlner, T., Niemier, M., Kogge, P.: Quantum-dot cellular automata. In: Electron Transport in Quantum Dots, pp. 397–431 (2003)
3. Nguyen, T.N., Liu, B.-H., Wang, S.-Y.: On new approaches of maximum weighted target coverage and sensor connectivity: hardness and approximation. IEEE Trans. Netw. Sci. Eng. **7**, 1736–1751 (2019)
4. Liu, B.-H., Nguyen, N.-T., Pham, V.-T., Lin, Y.-X.: Novel methods for energy charging and data collection in wireless rechargeable sensor networks. Int. J. Commun. Syst. **30**, e3050 (2017)
5. Babu, L., John, S.J., Parameshachari, B., Muruganantham, C., DivakaraMurthy, H.: Steganographic method for data hiding in audio signals with LSB & DCT. Int. J. Comput. Sci. Mob. Comput. **2**, 54–62 (2013)
6. Kianpour, M., Sabbaghi-Nadooshan, R.: A novel quantum-dot cellular automata X -bit × 32-bit SRAM. IEEE Trans. Very Large Scale Integr. (VLSI) Syst. **24**, 827–836 (2015)
7. Perri, S., Corsonello, P.: New methodology for the design of efficient binary addition circuits in QCA. IEEE Trans. Nanotechnol. **11**, 1192–1200 (2012)
8. Song, Z., Xie, G., Cheng, X., Wang, L., Zhang, Y.: An ultra-low cost multilayer RAM in quantum-dot cellular automata. IEEE Trans. Circuits Syst. II Express Briefs **67**, 3397–3401 (2020)
9. Kowsalya, T., Babu, R.G., Parameshachari, B., Nayyar, A., Mehmood, R.M.: Low area present cryptography in FPGA using TRNGPRNG key generation. CMC-Comput. Mater. Continua **68**, 1447–65 (2021)
10. Majeed, A.H., Alkaldy, E., Bin Zainal, M.S., Bin, M.D., Nor, D.: Synchronous counter design using novel level sensitive T-FF in QCA technology. J. Low Power Electron. Appl. **9**, 27 (2019)
11. Huang, J., Lombardi, F.: Design and test of digital circuits by quantum-dot cellular automata. Artech (2007)

12. Sen, B., Goswami, M., Mazumdar, S., Sikdar, B.K.: Towards modular design of reliable quantum-dot cellular automata logic circuit using multiplexers. Comput. Electr. Eng. **45**, 42–54 (2015)
13. Alharbi, M., Edwards, G., Stocker, R.: Novel ultra-energy-efficient reversible designs of sequential logic quantum-dot cellular automata flip-flop circuits. J. Supercomput. **79**(10), 11530–57 (2023)
14. Amirzadeh, Z., Gholami, M.: Asynchronous counter in QCA technology using novel D flip-flop. Eur. Phys. J. Plus **139**(4), 352 (2024)

Chapter 4
Quantum Natural Language Processing: Revolutionizing Language Processing

Anjusha Pimpalshende, Madhu Bala Myneni, and Sarat Chandra Nayak

Abstract Quantum computers bring a new level of computational power, making use of the quantum mechanical phenomenon to enable quicker and more efficient solutions to complicated issues. This chapter examines the fascinating intersection of quantum computing and natural language processing, which has generated increasing interest in the new discipline of Quantum Natural Language Processing. This hybrid topic covers a broad range of NLP activities and uses the capabilities of quantum mechanics to handle important language processing issues. This chapter offers a comprehensive review of the current landscape of Quantum Natural Language Processing, categorizing the approaches that have been developed so far into two main groups: theoretical research as well as hardware implementations, whether conventional or quantum. These methods are further divided into task-specific categories, which include specialist NLP applications such as question-answering and sentiment analysis, as well as general-purpose uses like syntax-semantic representation. The chapter also explores the benefits of QNLP, discussing its advantages in terms of performance and methodology. Quantum Natural Language Processing is still in its early stages, this overview serves as a foundation for identifying future research directions. Quantum computers are not designed to replace classical systems, but rather to tackle specific challenges that traditional computing methods struggle to address.

Keywords Quantum computing · Natural language processes · Quantum algorithms · Qubits

A. Pimpalshende (✉) · M. B. Myneni
Department of Computer Science and Engineering, VNR VJIET, Hyderabad, India
e-mail: anjusha_p@vnrvjiet.in

M. B. Myneni
e-mail: madhubala_m@vnrvjiet.in

S. C. Nayak
Department of Computer Science and Engineering, GITAM University, Hyderabad, India
e-mail: saratnayak234@gmail.com

S. Dehuri et al. (eds.), *Advances in Quantum Inspired Artificial Intelligence*, Intelligent Systems Reference Library 274, https://doi.org/10.1007/978-3-031-89905-8_4

4.1 Introduction

Over the past decade, quantum computers, emerged in the field of information processing. Information is processed in quantum computers using the concepts of quantum mechanics, utilizing values of 0, 1, or both simultaneously, known as qubits or quantum bits. This enables quantum computers to carry out multiple computations simultaneously and address issues easily than traditional computers [1]. Despite the significant computing power of current computer systems; many tasks in Artificial Intelligence (AI) remain nearly unsolvable with today's computational capacity. Due to their capacity to analyze several possible states at once, quantum computers may handle difficult or time-consuming problems significantly more quickly. Utilizing quantum computing, quantum AI improves machine learning algorithm performance and creates more robust AI models. While AI traditionally uses algorithms to make predictions, analyze data and automate tasks, its potential is constrained by the limitations of classical computers in processing complex data. Quantum AI offers the potential to overcome these limitations by addressing challenges related to data size, complexity, and the speed of problem-solving. The primary goal is to develop and implement algorithms that can run more efficiently on quantum hardware compared to classical computers. Shor's algorithm in quantum computing allows rapid factorization of prime numbers, offering exponential speedup over any known classical algorithm. A qubit is the basic unit of information in quantum computing. In classical computing, a bit exists in one of two states (either 0 or 1). A qubit can exist in a state of 0, 1 or a combination of both simultaneously. Dirac notation is used in quantum computing to express a qubit's potential states, denoted as $|0\rangle$ for the state 0 and $|1\rangle$ for the state 1. A qubit is in multiple states at the same time allow quantum computers to process tasks in parallel and solve some problems far more quickly than conventional computer.

A quantum mechanical principle known as superposition permits distinct elements to take on a variety of configurations, with the overall state being a composite of all of these potentialities. In natural language processing, the superposition may be able to address some frequent problems like polysemy and lexical ambiguity more effectively. For example, the word "bar" might mean numerous things. It may be used to identify a metal object or a location that serves alcoholic beverages. The Dirac notation can be used to represent the word bar as a superposition state: $|bar\rangle = a|place\rangle + b|rod\rangle$. Entanglement is a feature that distinguishes qubits from the conventional bit. Unlike traditional bits, which can hold only one value (0 or 1) at a time, quantum entanglement allows qubits to exist in multiple states simultaneously. Bell states are special quantum states that represent the simplest and most extreme examples of quantum entanglement. Entanglement is a fundamental property of quantum mechanics that enables powerful quantum phenomena and underpins many quantum computing applications. When two qubits are entangled, their states become intrinsically linked, even if they are separated. For instance, consider two qubits assigned to distinct elements, a and b and prepared in the Bell state $(1/\sqrt{2}(|00\rangle + |11\rangle)$, if a is measured as $|0\rangle$ then b must also be $|0\rangle$. This occurs because $|00\rangle|00\rangle\rangle$ is the only state in the superposition where a is $|0\rangle$. These two qubits are split up and allocated to a and b as two distinct entities. When the qubits of a are measured, the result is either $|0\rangle|0\rangle$ or $|1\rangle|1\rangle$. Due to the entanglement property, the qubits assigned of b must yield the same as those assigned to a. In NLP, the encoded words with the grammatical structures establish specific relationships between them and interact like entangled quantum states. Quantum measurement is the process of observing a qubit in superposition, causing it to collapse into one of its possible states. *Quantum interference* is a special property of qubits. It lets us adjust their superposition so they are more likely to end up in certain states when measured. This is useful because it helps guide the qubit toward the results we want in quantum computations. A key distinction between Quantum AI and Classical AI is how they process information. Classical AI operates using classical bits and are processed sequentially [2]. In contrast, Quantum AI leverages qubits, which, due to superposition and entanglement, can exist in multiple states at once. Due to their intrinsic parallelism, quantum computers can process enormous volumes of data at once, which could result in an exponential speedup over classical systems for particular jobs.

4.2 Natural Language Processing with Quantum Computing

The field of natural language processing, which combines computer science and artificial intelligence (AI), uses machine learning to help computers comprehend, analyze, and generate human language. Advances in generative AI have been greatly aided by NLP research, enabling large language models (LLMs) to develop sophisticated communication skills and allowing image generation models to interpret and respond to user requests effectively [3]. NLP is an integral part of daily life, such as driving search engines, chatbots, GPS systems. Additionally, NLP is increasingly used in enterprise solutions to streamline operations, enhance employee productivity, and automate business processes. By enabling communication in the natural language, NLP makes it easier for people to collaborate with machines. It can be used for many different applications, such as spam detection, text summarization, machine translation, and chabot development. Quantum algorithms for NLP have led to the emergence of a new area of research, known as quantum natural language processing. [3] QNLP focused on designing algorithms for natural language processing tasks with quantum mechanism. Researchers Mehrnoosh Sadrzadeh, Stephen Clark, and Bob Coecke of the University of Oxford are credited with pioneering it. Grammar algebra, a mathematical framework for representing grammar in natural languages, was developed by Mehrnoosh Sadrzadeh. Word embedding—a method of representing words as vectors in a space that encodes their meanings—was Stephen Clark's main area of interest. BobCoecke was an expert in categorical quantum mechanics, which uses decomposable processes to explain quantum physics. The DisCoCat Model: In 2010, the researchers proposed the Distributional Compositional Categorical (DisCoCat) model, which combines the strengths of these three areas. The model integrates embedded words: representing words as vectors to capture their meanings. Grammatical structure: Using algebra to combine word meanings according to the syntax of a sentence. The DisCoCat model encodes the meaning of sentences into quantum circuits, allowing for quantum computations to process language data. It is claimed that this approach achieves exponential speed-up compared to classical implementations [3, 4]. Quantum Natural Language Processing emerges as a promising approach to address the complex challenges of understanding and processing human language. Classical methods, while powerful, often struggle with efficiency, scalability, and contextual nuance. Here's how QNLP offers unique advantages: Handling High Dimensionality: Natural language involves vast amounts of data that require high-dimensional representations, such as word embeddings and sentence encodings. Quantum computers, with their ability to process exponentially large state spaces, manage these high-dimensional structures more efficiently than classical systems. Semantic representation.

Human language is rich with context and relationships between words, sentences, and phrases. Principles of quantum physics include entanglement and superposition are particularly suited for capturing these intricate semantic relationships, enabling richer and more nuanced representations of meaning. Because language is compositional by nature, a sentence's meaning is determined by the meanings of its constituent words. Quantum frameworks, like categorical quantum mechanics, naturally model these compositional structures. This approach enhances tasks like translation, summarization, and grammar parsing by faithfully representing how meanings combine. Efficient Processing: Classical NLP models, particularly deep learning approaches, require massive computational resources for training and inference. Quantum computing could potentially solve some NLP problems—such as parsing, sentiment analysis, or syntax tree generation—faster and with lower resource consumption, offering significant efficiency gains. Pattern Recognition and Ambiguity Resolution: Pattern recognition and ambiguity (e.g., resolving word meanings in context) are core challenges in NLP. Quantum computers, with their ability to analyze and process patterns in parallel, excel in tasks like: Entity recognition, Coreference resolution, Word-sense disambiguation. Optimization for Large Models Training (LLMs) large language models is computationally intensive and expensive. Quantum-enhanced optimization techniques could streamline this process, enabling faster and more cost-effective training of advanced NLP systems. Advancing AI Research: QNLP is a convergence of quantum computing and natural language processing, pushing the boundaries of both fields. It has the potential to reveal entirely new algorithms, paradigms, and methods that are infeasible with classical computation alone. By leveraging the unique capabilities of quantum mechanics, QNLP promises to handle the inherent complexities of language more effectively, opening doors to advanced AI systems that are faster, more efficient, and more capable of understanding the subtleties of human communication.

4.3 Quantum Algorithms and Quantum Approach for NLP Applications

To tackle Natural Language Processing (NLP) problems more quickly, Quantum algorithms make use of the special powers of quantum computers, including parallel processing, entanglement, and superposition. Some important quantum algorithms for NLP are Shor's algorithm [4] and Grover's algorithm [3]. It demonstrates the power of quantum computing. Classical computers would take billions of years to solve the prime factorization problem, but Shor's algorithm can do it in just a few hours. Grover's algorithm helps to find an item in an unsorted database much faster than classical methods, offering a significant speed boost.

These algorithms are explained in more detail below.

Shor's algorithm was invented by Peter Shor and runs in much less time and space compared to classical methods. The goal is to find a number p, between 1 and N, that evenly divides N to quickly find the factors of a number N. Shor's algorithm has two main steps: Reducing the problem—It first changes the factoring problem into a different problem called "order-finding," which can be solved using a regular (classical) computer. Using quantum computing—A quantum algorithm is then used to solve the "period-finding" problem efficiently.

The procedure is explained in Algorithm 1.

Algorithm 1: Shor's algorithm

Classical part:

1. Choose an arbitrary number a ($a< N$).
2. Use the Euclidean technique to compute gcd(a, N).
3. The procedure is finished if gcd (a, N) $\neq 1$, indicating that N has a nontrivial factor.
4. Determine r, the period of the function $f(x)=a$ x mod N, using the period-finding subroutine if gcd (a, N) = 1. The smallest integer r that makes $f(x + r) = f(x)$ must be found.
5. Return to step 1 if r is odd.
6. Go back to step 1 and select a new n if $a^{r/2}=-1$ mod N.
7. gcd ($a^{r/2}\pm 1$, N) is the factor of N. The procedure has finished.
8. Quantum part: Use Period-finding subroutine

Grover's algorithm: is a quantum method for unstructured search problems, it offers a quadratic improvement over classical algorithm. Example: Search a phone book for a specific number, travelling salesman problem (TSP). Classical search takes O(N) steps and Grovers' algorithm takes O ($\sqrt{N}$) steps. In NLP, Grover's algorithm can be applied to text retrieval, information extraction, keyword-based search, similarity detection, Named Entity Recognition (NER),

Grover's Algorithm for NLP Applications

Step 1: Problem Setup:

- Identify the target f(x) =1, which indicates a "match" (e.g., finding a sentence containing a specific word, a document containing a named entity, or the most similar sentence to a query).
- All other elements x in the dataset have f(x)=0.

Oracle Function:

- Create a quantum oracle Zf that flips the amplitude of the target state (e.g., a sentence or word satisfying f(x)=1).
- For NLP, the oracle can be based on:
  - String matching.
  - Semantic similarity scoring.
  - Named entity recognition (NER).

Step 2: Initialize the Superposition

1. Represent the dataset (e.g., N possible words, sentences, or documents) as a quantum state of n-qubits, where N = 2^n.
2. Initialize the quantum register Q in the $|0^n\rangle$ state.
3. Apply the Hadamard operation to create an equal superposition over all possible candidates:

$$|\psi\rangle = \frac{1}{\sqrt{N}} \sum_{x=0}^{N-1} |x\rangle$$

Step 3: Apply Grover Iterations

Repeat the following steps t times ($t \approx \frac{\pi}{4}\sqrt{N/M}$)(where M is the number of solutions):

1. Use the Oracle (Zf) first:
   - Flip the amplitude of the target states identified by f(x) =1.
   - For NLP, this involves:
     - Checking if a word or phrase matches the query.
     - Calculating semantic similarity and marking states above a threshold.
     - Identifying entities using quantum-encoded rules.
2. Use the ZOR (Diffusion Operator):
   - Reflect the quantum state around the average amplitude to increase the amplitude of the indicated states:

$$Z_{OR} = 2|\psi\rangle\langle\psi| - I$$

   - This increases the likelihood of measuring the correct solution.

Step 4: Measurement

1. After t repetitions, measure the quantum state in the computational basis.
2. The output is the index (or indices) of the most probable states, which correspond to the solution(s).

Some more algorithms used in NLP applications are Quantum Walks [5] for Parsing to model sentence structure and syntactic parsing. Example: Speeding up dependency parsing or constituency parsing. Quantum Support Vector Machines

(QSVM) [6] to Perform text classification tasks more efficiently using quantum-enhanced SVMs. Quantum Neural Networks (QNN) [7] to train quantum-inspired neural networks for NLP tasks. Example is developing quantum transformers for sequence-to-sequence tasks like translation. Quantum Natural Language Models (QNLMs) [8] to implement models specifically designed for processing natural language using quantum principles. Example: Quantum compositional models for sentence meaning. Quantum-Enhanced Optimization [9] to optimize large-scale NLP models. Example: Fine-tuning large language models like GPT more efficiently. Quantum Algorithms for Semantic Similarity [10] to Measure the semantic similarity between words, phrases, or documents. Quantum Sequence Modeling to model sequential data like text using quantum systems. Example is Quantum-enhanced chatbot or language translation systems. Quantum Machine Translation [11] to improve efficiency and accuracy of translating between languages. Example is Quantum machine translation for low-resource languages. Textual data can be mapped to high-dimensional quantum feature spaces using quantum kernel methods for tasks like regression and classification. Text classification (such as spam detection) is one of the applications. Sentiment analysis with word embeddings and quantum kernels.

4.4 Challenges in Quantum NLP

While achieving quantum supremacy is a significant focus, challenges persist. Quantum computers are highly sensitive to external influences such as temperature and electromagnetic radiation. Ensuring the stability of qubits continues to be a major technical hurdle. Building scalable and error-tolerant quantum computers is an ongoing research area. Quantum systems are prone to errors. This makes developing error-correction methods essential for making quantum computing and Quantum AI reliable and practical in real-world use. Hardware Restrictions: Since quantum computers are still in their infancy, their error rates and qubit counts are limited. Algorithm Development: A lot of NLP quantum algorithms are still in the theory stage and require extensive testing and development. Integration with Classical Systems: It is difficult to combine quantum and classical parts in an efficient way. It is really a challenge.

4.5 Quantum NLP Applications

Quantum computing has the potential to transform natural language processing (NLP) by offering computational advantages in areas where classical methods struggle due to the complexity and scale of the data. As quantum hardware improves

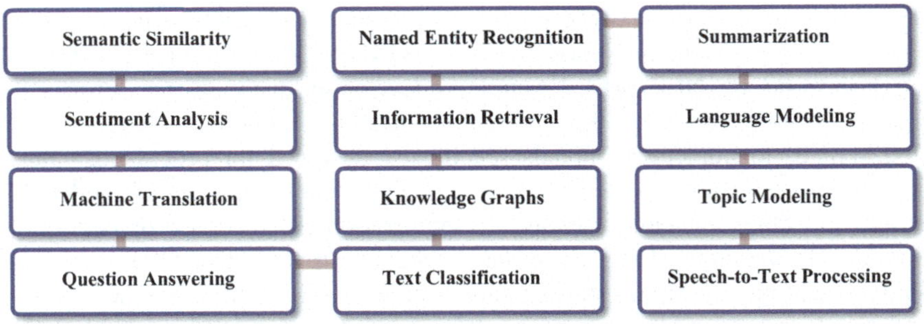

Fig. 4.1 Landscape of QNLP applications

and hybrid quantum–classical approaches mature, NLP could benefit from breakthroughs like real-time translation, deeper semantic analysis, and more human-like conversational agents. Figure 4.1 gives landscape of QNLP applications.

Following table (Table 4.1) gives idea about NLP applications with Quantum computing.

4.5.1 Semantic Similarity

Semantic similarity is how closely two words, sentences, or documents are related in meaning. QNLP can leverage quantum mechanics principles and quantum computing to model semantic similarity more efficiently. The notable concepts included are.

Distributional Semantics Words and phrases are represented in a high-dimensional vector space called embeddings. These vectors are encoded called quantum states (qubits) by leveraging the **Hilbert space** representation.

Quantum Entanglement and Superposition semantic meaning often arises from word combinations. Quantum superposition and entanglement model these interactions, capturing relationships better than classical embeddings.

Quantum measurement Quantum measurement determines how similar two quantum states are by computing the overlap or fidelity between them.

Categorical Compositional Models (DisCoCat) is a quantum-inspired framework where grammatical structures (syntax) guide how meanings (semantics) combine.

Cross-lingual research in language processing adopted word embeddings to generate an atlas for 76 different languages [12]. This knowledge helps for applications like sentiment analysis. A query-based framework has been developed with specific guidelines for user interface and understanding, information retrieval, and semantic web [13]. The semantic structure has been restructured with feature weights to analyze subject-action-object (SAO), which helps to build professional vocabulary for measuring technology similarity used in research on Alzheimer's disease [14]. Heterogeneous knowledge graphs are used in the domain of education to

Table 4.1 Summary of QNLP applications with algorithms used

Application	Description	Quantum algorithms
Semantic similarity	Enhancing the identification of similarities between words, phrases, or documents using quantum principles	Quantum Kernel Methods, Quantum Annealing
Sentiment analysis	Leveraging quantum computing to process large datasets for faster and more nuanced sentiment detection	Quantum Support Vector Machines (QSVM), Quantum Feature Maps
Machine translation	Using quantum algorithms to speed up translation tasks across multiple languages	Quantum Approximate Optimization Algorithm (QAOA)
Question answering	Employing quantum techniques to enhance retrieval and interpretation of answers from large datasets	Grover's Search, Quantum Walks
Text classification	Optimizing categorization tasks such as spam detection, topic tagging, and document sorting	Variational Quantum Circuits (VQC), Quantum Neural Networks (QNN)
Knowledge graphs	Using quantum mechanics-inspired models for building and querying large-scale semantic graphs	Quantum Graph Neural Networks (QGNNs), Quantum Annealing for Graph Embedding
Information retrieval	Enhancing search engines by leveraging quantum-inspired algorithms to rank and retrieve documents	Amplitude Amplification, Grover's Search
Named entity recognition	Using quantum models for recognizing names, dates, locations, etc., from unstructured text	Quantum Hidden Markov Models (QHMM), Quantum Variational Circuits
Summarization	Applying quantum approaches for both extractive and abstractive summarization tasks	Quantum Principal Component Analysis (QPCA), Quantum Annealing
Language modeling	Building quantum-based models to predict next words in a sequence or understand text context	Quantum Boltzmann Machines, Quantum Annealing
Topic modeling	Quantum algorithms for clustering and identifying hidden themes in large text corpora	Quantum Latent Dirichlet Allocation (QLDA)
Speech-to-text processing	Quantum methods for improving the efficiency of converting audio speech to text	Quantum Fourier Transform (QFT), Quantum Phase Estimation

(continued)

Table 4.1 (continued)

Application	Description	Quantum algorithms
Cross-lingual NLP	Using quantum computing for tasks involving multiple languages, such as code-switching and multilingual retrieval	Quantum Embedding Techniques, Quantum Neural Networks
Adversarial text generation	Leveraging quantum techniques to generate and detect adversarial examples in NLP models	Quantum Generative Adversarial Networks (QGANs), Quantum Annealing for Robustness Testing

address the mapping of learned courses and job market requirements [15]. The next era of research focus is on the quantum semantic communication framework [16]. It aids in extracting relevant information from data and encoding it into compact high-dimensional quantum states for precise communication across quantum networks.

The sentence representation uses the tensor product to entangle two successive words and quantum entanglement-triggered dependencies words [17]. The research focus is on algorithm development to address the problems in a bipartite quantum system [18], focused on entangled subspaces of matrices having no product state and decomposition of a tensor. Computing sentence similarity [19] Using quantum entanglement helps reduce semantic noise by minimizing the impact of the entangled word vector's tensor product. Using an improved (QPSO) Quantum Particle Swarm Optimization, a new word-level adversarial assault technique was presented [20]. It improves the effectiveness of adversarial attacks in language processing models.

4.5.2 Sentiment Analysis

It is a tool to understand the conceptual meaning of large documents. By representing words, sentences, and their semantic relationships as quantum states and operations, QNLP enables the handling of large-scale text data while exploring complex contextual interactions. The performance of sentiment analysis in natural language processing is enhanced by the combination of quantum states, superposition, and entanglement. The key contributions in QNLP for sentiment analysis:

- **Quantum Encoding**: Representing words and their relationships using quantum states allows capturing rich semantic information with higher-dimensional structures.
- **Contextual Understanding**: QNLP uses quantum circuits to model sentence structures, enabling better interpretation of context and emotional tone.
- **Improved Scalability**: Quantum systems offer faster processing for large datasets compared to classical NLP models.

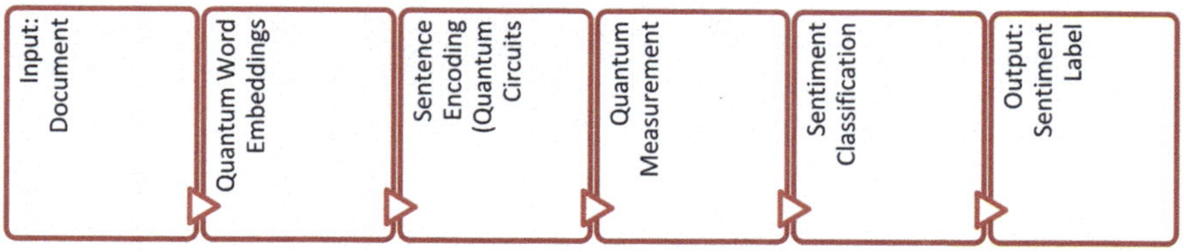

Fig. 4.2 Variational quantum algorithms framework

- **Efficient Feature Extraction**: Quantum entanglement helps in capturing interdependent features for sentiment classification.
- **Applications**: Real-time sentiment monitoring in social media, customer feedback analysis, and opinion mining across industries (Fig. 4.2).

Variational Quantum Algorithms framework [21, 22] was utilized for quantum sentiment analysis, emotion, and sarcasm classification. It offers optimization, metric measurement, diagram-to-circuit conversion, and sentence-to-diagram conversion. The Quantum-Enhanced Support Vector Machine [23] method has been used to analyze the sentiment. The pipeline consists of creating state vectors, training circuit parameters, converting sentences to circuits, and then training and evaluating the model. A multi-modal framework made up of sarcasm, sentiment, and emotion analysis tasks has been proposed as: quantum probability driven multi-modal sarcasm, sentiment, and emotion analysis [24]. The emotion has been analyzed Quantum state preparation is a key step in sentiment analysis, which is implemented using deep Qprep [19]. A Quantum-Enhanced Multi-Modal Sentiment Analysis [25] framework was evaluated under the environment of IBM's Qiskiton CMUMOSEI dataset [26].

4.5.3 Machine Translation

Machine translation offers a clear way to understand how text is converted from one language to another. The key components are quantum embeddings, quantum circuit generation, Quantum superposition, Quantum entanglement for contextual understanding, Quantum algorithms like Variational Quantum Circuits (VQC) optimize the translation process by fine-tuning model parameters for better contextual accuracy and Quantum Approximate Optimization Algorithm (QAOA) improve alignment between source and target languages (Fig. 4.3).

Machine translation from English to Persian at the sentence level is made possible by cross-language translation employing quantum circuits with Q-LSTM cells. [27, 28]. Quantum translation model was developed for Chinese to English language translation [29].

The primary areas where quantum computing is being investigated for NLP applications are:

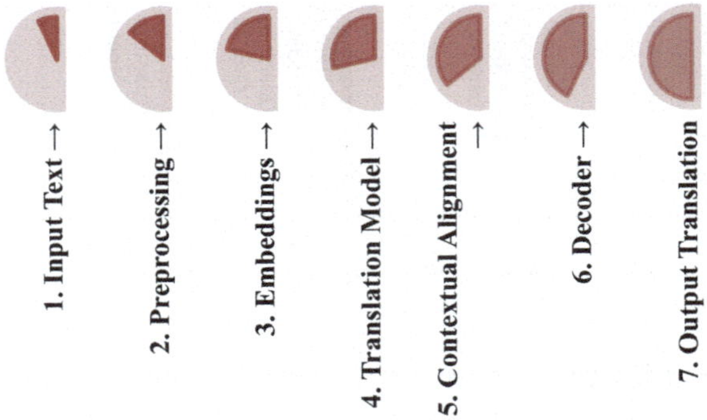

Fig. 4.3 Quantum machine translation key components

1. **Efficient Optimization**
    - Quantum Annealing: Quantum algorithms like quantum annealing can optimize large-scale NLP models, such as transformer architectures, by finding better solutions for complex objective functions.
    - Parameter Optimization: Quantum optimization can improve hyperparameter tuning in NLP models, which is critical for achieving high performance.
2. **Improved Embedding Representations**
    - Quantum Word Embeddings: Quantum states can encode complex relationships and semantic similarities between words in fewer dimensions, potentially outperforming classical embeddings like Word2Vec, GloVe, or BERT.
    - Entanglement and Superposition: These principles can represent multiple word meanings simultaneously, providing richer context-aware representations.
3. **Speeding Up Matrix Operations**
    - Quantum Linear Algebra: Many NLP tasks involve operations like matrix multiplications, which are computationally intensive. Quantum algorithms, like the Harrow-Hassidim-Lloyd (HHL) algorithm, can speed up these operations. Transformers and Attention Mechanisms: Quantum-enhanced methods enable the processing of longer sequences by reducing the computational cost of attention mechanisms in transformers.
4. **Probabilistic and Generative Models**
    - Quantum Probabilistic Models: Quantum computers excel at simulating probabilistic systems. They could enhance models like Conditional Random Fields and Hidden Markov Models, which are frequently employed in NLP.
    - Quantum Generative Models: Variants of quantum-enhanced generative adversarial networks (QGANs) could improve text generation tasks.

5. **Semantic Analysis**
   - Quantum Logic: Quantum computing can handle the ambiguity and fuzziness in natural language more effectively using quantum logic, which mimics the probabilistic nature of human reasoning.
   - Semantic Parsing: Quantum circuits can process semantic structures with higher efficiency.
6. **Parallelism and Large-Scale Data Handling**
   - Quantum Parallelism: Quantum computers can explore several possibilities concurrently, which is beneficial for tasks like parsing and machine translation.
   - Handling Large Datasets: Quantum-enhanced clustering and dimensionality reduction techniques can handle the massive data scale in NLP tasks more efficiently than classical methods.
7. **Improved Language Modeling**
   - Quantum-enhanced Transformers: By integrating quantum components, transformer models like GPT and BERT might achieve improved efficiency and accuracy.
   - Contextual Understanding: Quantum principles can enable better disambiguation and deeper contextual understanding.

4.6 Conclusion

A new area called QNLP (quantum natural language processing) uses the ideas of quantum mechanics to more effectively describe linguistic structures than traditional methods. By encoding words, sentences, and grammatical relationships as quantum states in a Hilbert space, QNLP exploits superposition and entanglement to capture the intricate dependencies inherent in language. Quantum computing and natural language understanding, offering innovative solutions to challenges that have long constrained classical NLP. This approach enables more context-aware and computationally efficient NLP models, with potential applications in machine translation, question-answering systems, and semantic analysis. QML incorporates quantum computing into classical machine learning methods to improve performance and scalability, is closely related to QNLP. QML applies quantum principles to optimize models, improve training efficiency, and enable better handling of high-dimensional data. Key techniques in QML include: Quantum Data Encoding—Classical data, such as text or images, is mapped into quantum states for efficient processing. Quantum Neural Networks (QNNs): By simulating classical neural networks, quantum circuits enable learning tasks to be completed at potentially exponential speeds. Quantum Support Vector Machines (QSVMs): These quantum-enhanced SVMs are more effective at classifying complicated datasets. Quantum Boltzmann Machines—Quantum implementations of energy-based probabilistic models that improve optimization and sampling. The synergy between QNLP and QML could lead to groundbreaking advancements in AI, enabling quantum-enhanced NLP models that outperform classical deep learning approaches in both accuracy and efficiency. However, real-world applications are still in their infancy, and issues like algorithm design, error correction, and hardware constraints require more study. As quantum computing progresses, QNLP and QML hold immense potential for transforming artificial intelligence, leading to more sophisticated, scalable, and efficient language processing and learning systems. QNLP can model linguistic complexities more efficiently and open pathways to solving problems previously considered intractable. While still in its infancy, the promise of QNLP is immense, with the potential to redefine how we interact with and interpret language across diverse domains. The road ahead involves addressing technical hurdles, such as advancing quantum hardware and refining algorithms, but the possibilities are both exciting and transformative. As the field evolves, QNLP may well become a cornerstone of future advancements in artificial intelligence, expanding the limits of what can be achieved in language processing.

References

1. https://www.ibm.com/think/topics/quantum-computing
2. Nofer, M., Bauer, K., Hinz, O. et al.: Quantum computing. Bus. Inf. Syst. Eng. **65**, 361–367 (2023). https://doi.org/10.1007/s12599-023-00823-w
3. Du, S.L., Santana, S.H., Scarpa, G.: A gentle introduction to quantum natural language processing (2022). arXiv:2202.11766
4. Shor, P.W.: Polynomial-time algorithms for prime factorization and discrete logarithms on a quantum computer. SIAM J. Comput. **26**(5), 1484–1509 (1997). https://doi.org/10.1137/s0097539795293172
5. Venegas-Andraca, S.E.: Quantum walks: a comprehensive review. Quantum Inf. Process. **11**, 1015–1106 (2012). https://doi.org/10.1007/s11128-012-0432-5
6. Innan, N., Khan, M., Panda, B., et al.: Enhancing quantum support vector machines through variational kernel training. Quantum Inf. Process. **22**, 374 (2023). https://doi.org/10.1007/s11128-023-04138-3
7. Jeswal, S.K., Chakraverty, S.: Recent developments and applications in quantum neural network: a review. Arch. Comput. Methods Eng. **26**, 793–807 (2019). https://doi.org/10.1007/s11831-018-9269-0
8. Widdows, D., Aboumrad, W., Kim, D. et al.: Quantum Natural Language Processing. Künstl Intell (2024). https://doi.org/10.1007/s13218-024-00861-w
9. Cai, P., Shen, K., Yang, T., et al.: Enhancing quantum approximate optimization with CNN-CVaR integration. Quantum Inf. Process. **24**, 37 (2025). https://doi.org/10.1007/s11128-025-04655-3
10. Yu, K., Guo, G.D., Li, J., et al.: Quantum algorithms for similarity measurement based on euclidean distance. Int. J. Theor. Phys. **59**, 3134–3144 (2020). https://doi.org/10.1007/s10773-020-04567-1
11. Pandey, S., Basisth, N.J., Sachan, T., Kumari, N., Pakray, P.: Quantum machine learning for natural language processing application. Physica A: Stat. Mech. Appl. **627**, 129123 (2023). ISSN:0378-4371. https://doi.org/10.1016/j.physa.2023.129123
12. Şenel, L.K., Utlu, İ., Yücesoy, V., Koç, A., Çukur, T.: Generating semantic similarity atlas for natural languages. In: 2018 IEEE Spoken Language Technology Workshop (SLT), pp. 795–799. Athens, Greece (2018). https://doi.org/10.1109/SLT.2018.8639521

13. Houssein, E.H., Ibrahem, N., Zaki, A.M., Sayed, A.: Semantic protocol and resource description framework query language: a comprehensive review. Mathematics **10**(17), 3203 (2022)
14. Li, R., Wang, X., Liu, Y., Zhang, S.: Improved technology similarity measurement in the medical field based on subject-action-object semantic structure: a case study of Alzheimer's Disease. IEEE Trans. Eng. Manag. **70**(1), 280–293 (2023). https://doi.org/10.1109/TEM.2020.3047370
15. Ramazanova, V., Sambetbayeva, M., Serikbayeva, S., Sadirmekova, Z., Yerimbetova, A.: Development of a knowledge graph-based model for recommending MOOCs to supplement university educational programs in line with employer requirements. IEEE Access **12**, 193313–193331 (2024). https://doi.org/10.1109/ACCESS.2024.3519263
16. Chehimi, M., Chaccour, C., Thomas, C.K., Saad, W.: Quantum semantic communications for resource-efficient quantum networking. IEEE Commun. Lett. **28**(4), 803–807 (2024). https://doi.org/10.1109/LCOMM.2024.3361852
17. Yu, Y., Qiu, D., Yan, R.: An efficient framework for sentence similarity inspired by quantum computing. In: 2021 IEEE International Conference on Big Knowledge (ICBK), pp. 157–163. Auckland, New Zealand (2021). https://doi.org/10.1109/ICKG52313.2021.00030
18. Johnston, N., Lovitz, B., Vijayaraghavan, A.: Computing linear sections of varieties: quantum entanglement, tensor decompositions and beyond. In: 2023 IEEE 64th Annual Symposium on Foundations of Computer Science (FOCS), pp. 1316–1336. Santa Cruz, CA, USA (2023). https://doi.org/10.1109/FOCS57990.2023.00079
19. Yu, Y., Qiu, D., Yan, R.: A quantum entanglement-based approach for computing sentence similarity. IEEE Access **8**, 174265–174278 (2020). https://doi.org/10.1109/ACCESS.2020.3025958
20. Chen, Q., Sun, J., Palade, V.: A word-level adversarial attack method based on sememes and an improved quantum-behaved particle swarm optimization. IEEE Trans. Neural Netw. Learn. Syst. **35**(11), 15210–15221 (2024). https://doi.org/10.1109/TNNLS.2023.3283308
21. Ruskanda, F.Z., Abiwardani, M.R., Syafalni, I., Larasati, H.T., Mulyawan, R.: Simple sentiment analysis ansatz for sentiment classification in quantum natural language processing. IEEE Access **11**, 120612–120627 (2023). https://doi.org/10.1109/ACCESS.2023.3327873
22. Ruskanda, F.Z., Abiwardani, M.R., Mulyawan, R., Syafalni, I., Larasati, H.T.: Quantum-enhanced support vector machine for sentiment classification. IEEE Access **11**, 87520–87532 (2023). https://doi.org/10.1109/ACCESS.2023.3304990
23. Liu, Y., Zhang, Y., Song, D.: A quantum probability driven framework for joint multi-modal sarcasm, sentiment and emotion analysis. IEEE Trans. Affect. Comput. **15**(1), 326–341 (2024). https://doi.org/10.1109/TAFFC.2023.3279145
24. Phukan, A., Pal, S., Ekbal, A.: Hybrid quantum-classical neural network for multimodal multitask sarcasm, emotion, and sentiment analysis. IEEE Trans. Comput. Soc. Syst. **11**(5), 5740–5750 (2024). https://doi.org/10.1109/TCSS.2024.3388016
25. Selig, P., Murphy, N., Redmond, D., Caton, S.: DeepQPrep: neural network augmented search for quantum state preparation. IEEE Access **11**, 76388–76402 (2023). https://doi.org/10.1109/ACCESS.2023.3296802
26. Singh, J., Ali, F., Shah, B., Bhangu, K.S., Kwak, D.: Emotion quantification using variational quantum state fidelity estimation. IEEE Access **10**, 115108–115119 (2022). https://doi.org/10.1109/ACCESS.2022.3216890
27. Phukan, A., Asif, E.: QeMMA: Quantum-Enhanced Multi-Modal Sentiment Analysis. ICON (2023)
28. Abbaszade, M., Salari, V., Mousavi, S.S., Zomorodi, M., Zhou, X.: Application of quantum natural language processing for language translation. IEEE Access **9**, 130434–130448 (2021). https://doi.org/10.1109/ACCESS.2021.3108768
29. Vydeki, D., Victor, D., Mamoo, Z., Rohit, V., Annis Fathima, A.: Machine translation using dictionary based techniques. In: 2023 IEEE 8th International Conference for Convergence in Technology (I2CT), pp. 1–6. Lonavla, India (2023). https://doi.org/10.1109/I2CT57861.2023.10126256

Chapter 5
Quantum Machine Learning in Climate Science and Climate Change Solutions

Barsa Priyadarshani Behera and Monalisa Jena

Abstract Climate change represents one of the most significant challenges faced today, requiring innovative and efficient strategies to mitigate its effects and adapt to changing environmental circumstances. In response, quantum machine learning (QML) has surfaced as an exciting interdisciplinary domain. By leveraging the computational benefits of quantum computing along with the predictive capabilities of machine learning, QML presents encouraging novel methods for addressing intricate climate-related issues. Climate science, with its complex, high-dimensional data and urgent need for actionable insights, presents a promising application domain for QML. QML has the potential to change climate science by making data analysis faster, improving climate models, and giving better predictions about climate change. QML tries to find new solutions to big climate problems like risk assessment, disaster planning, and reducing carbon emissions. This chapter provides a comprehensive overview of climate science fundamentals, reviews major QML techniques applied in climate science, and explores the synergies between quantum computing and machine learning in advancing sustainable climate solutions.

Keywords Quantum computing · Machine learning · Climate science · Climate change · Quantum machine learning · Artificial intelligence

5.1 Introduction

In recent decades the Earth's average temperature has increased rapidly, rising sea levels, leading to glacier melting, and more frequent extreme weather events. This is due to burning fossil fuels like coal, oil, and gas, and cutting trees. Today, the global temperature is about 1 °C higher than before industrialization or pre-industrial times and scientists warn that exceeding 1.5 °C could trigger irreversible "tipping points," causing permanent changes to a hotter climate [1]. While climate change is a global issue, its effects vary by region, and understanding future climate changes from

B. P. Behera (✉) · M. Jena
Department of Computer Science, Fakir Mohan University, Balsore, Odisha, India
e-mail: barsabehera600@gmail.com

S. Dehuri et al. (eds.), *Advances in Quantum Inspired Artificial Intelligence*, Intelligent Systems Reference Library 274, https://doi.org/10.1007/978-3-031-89905-8_5

regional levels to local is essential for effective policies [2]. However, this remains a major challenge with massive amounts of data from climate models, sensors, and satellites. The uncertainty in datasets includes errors and data gaps that occur during acquisition, storage, and transmission [3]. The complexity of the climate system and its representation makes it difficult to accurately predict responses to changes such as shifts in atmospheric transparency, variations in sunlight reaching Earth, or the movement of continents.

Quantum computing combines ideas from computer science, information theory, and quantum physics. It uses quantum mechanics to process information by studying tiny particles like photons and electrons. Quantum computers are much faster and more efficient than classical computers, solving complex problems that would take traditional systems thousands of years. They have applications in finance, supply chains, climate science, healthcare, AI, and cybersecurity. Some fields, like machine learning and chemistry, are already benefiting from quantum technology. Artificial intelligence (AI) and machine learning (ML) help overcome these challenges by quickly and accurately analyzing big datasets, finding patterns, and making predictions. It is also used for climate change mitigation, including carbon capture, smart building systems, better transportation, and efficient waste management [4].

In the past five years, AI & ML have been the most popular topics among researchers. Some well-known AI methods are closely related to how biological brains work or how the human brain processes information [5]. Advanced AI algorithms equally need strong hardware progress, and quantum computing offers a promising solution [6]. With better computers and smarter algorithms, ML has become a powerful way to find patterns in data [7, 8]. While AI and ML are the focus now, quantum computing is expected to lead future research trends. AI and quantum computing share some similarities. While machine learning enables computers to solve problems by learning from data or experience without explicit programming, quantum computing processes information based on the principles of quantum theory [9, 10].

Classical computers need significant time and resources to handle such tasks. Quantum computers, using qubits can represent 0 and 1 simultaneously through superposition and entanglement, process large data more efficiently [11, 12]. This makes quantum computing ideal and powerful for ML. Quantum mechanics is known for producing unique patterns and offers new possibilities for machine learning. The new concept called quantum intelligence (QI) is the combination of AI and quantum computing [13]. It looks at how combining these two technologies can solve the complex problems that regular computers cannot handle [14]. The basic quantum algorithms include quantum versions of classical machine learning methods like support vector machines and k-nearest neighbor models, for example; unsupervised learning has been used to detect quantum entanglement in different structures [15]. Similarly, deep learning methods like quantum neural networks are major applications in image classification, help reduce noise in quantum systems, and analyze molecular structures and dynamics. The applications in industries are; finance (risk analysis and fraud detection), healthcare and pharmaceuticals, supply chain and logistics, cybersecurity, energy, automotive, telecommunications, retail, aerospace, climate science, entertainment and media, and government and defense.

The chapter is organized as follows: In Sect. 5.2, the basics of climate science are discussed. Section 5.3 presents quantum AI and ML techniques along with their applications in climate science. Section 5.4 explores the role of quantum machine learning in climate science, including relevant mathematical equations and explanations. Lastly, Sect. 5.5 concludes the chapter with insights on potential future research directions.

5.2 Climate Science Basics

Climate science studies the Earth's climate system, its changes, and its impacts. It integrates knowledge from meteorology, oceanography, physics, and chemistry to understand climate patterns. It also helps us understand how the Earth's climate works, how it changes over time, and how human activities influence it. To mitigate climate-related risks, climate science is necessary for predicting future climate conditionsand developing strategies. The twentieth century saw major advancements with the development of climate models and satellite observations. Today, climate science is a critical area of global concern.

5.2.1 Components of Climate Science

Temperature, precipitation, wind, and humidity are the key elements of climate. The climate system consists of several interconnected components that interact to regulate Earth's climate such as the atmosphere (the gaseous layer surrounding Earth that controls temperature, precipitation, and wind patterns), hydrosphere (water bodieslikeoceans, lakes, and riversregulate temperature and store heat, significantly influencing climate), cryosphere (ice and snow reflect sunlight, affecting global temperatures and sea levels), biosphere (living organisms' role in climate such as processes like respiration, photosynthesis, and carbon storage), and lithosphere (land and geological factors).

5.2.2 Climate Change and Its Causes

The persistent change in the Earth's temperature, weather patterns, and overall climate over a long period is called climate change. which can occur naturallythrough changes in solar radiation, volcanic activity, ocean circulation, or human activities like greenhouse gas emissions (burning fossil fuels), deforestation, industrialization, and urbanization.

5.2.2.1 Climate Change Effects

Increasing temperatures are responsible for heatwaves, ice melt, sea levels rising, coastal flooding, shifting climate zones, and higher energy demands. Extreme weather events or more frequent and intense natural disasters like hurricanes, droughts, and floods threaten communities and economies.

5.2.2.2 Climate Change Impacts

The impact on ecosystems and biodiversity results in species migration and extinction, disrupting natural habitats and altering ecological balance. Climate change forces species to relocate in search of suitable conditions, while others struggle to adapt, leading to biodiversity loss and ecosystem instability. Changing rainfall patterns affect crop yields and food production systems; this is a huge impact on agriculture and food security. Socio-economic impacts like economic losses, health issues, and displacement of communities due to climate-related disasters [2, 3].

5.2.3 Climate Modelling and Prediction

Climate models are essential tools for understanding climate patterns and making future predictions, enabling policymakers to make informed decisions to combat climate change. Climate data collection and analysis are done by satellite observations that provide global climate data and weather stations orsensors that collect ground-based climate information. The climate models are: general circulation models (GCMs) that simulate global climate processes, Earth system models (ESMs) which are the latest generation of climate models, and regional climate models (RCMs) which focus on specific areas for detailed analysis.

5.2.4 Climate Science and Technology

In climate science, remote sensing technologies like satellites and drones provide valuable climate observations, AI helps mainlyanalyzeclimate data and improve predictions, quantum computing enhances complex climate simulations, and big data aids in better decision-making and trend analysis.

5.3 Major Quantum AI & ML Techniques Used in Climate Science

QML involves using quantum principles or technology to enhance machine learning and applying machine learning techniques to improve quantum computing. It mainly depends on two key factors: data and algorithms. They can be either quantum or classical as shown in Fig. 5.1 [16]. There are different types of QML methods based on the type of data and the type of computer used. It can be started with classical data (traditional numerical data) and quantum data (data from quantum systems). Depending on the data type, it is processed using either a classical computer or a quantum computer. This can lead to four types of machine learning approaches. Classical machine learning is the standard method which uses classical computers to process classical data, quantum-applied machine learning still uses classical computers but applies some quantum techniques to improve performance, quantum-enhanced models use quantum computers to process classical data more efficiently, and the generalized quantum machine learning is fully quantum-based, using quantum computers for both quantum data and computations.

This section exploresseveral quantum AI and ML techniques used in climate science for various computational tasks.

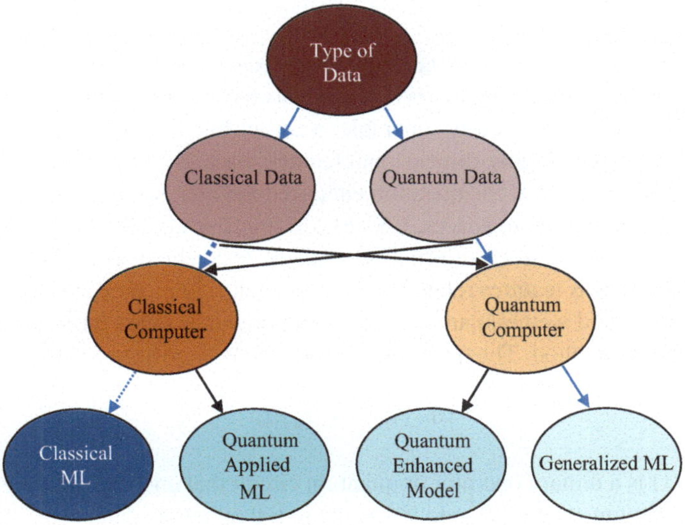

Fig. 5.1 Quantum machine learning paradigms

5.3.1 Quantum Support Vector Machine (QSVM)

The SVM is a widely used classifier used mainly for binary classification tasks. It operates by finding the hyperplane that best separates data points into two classes in the feature space. The data points closest to the hyperplane are referred to as support vectors. SVM aims to obtain the maximum margin between the support vectors [17]. SVM also excels handling data in high-dimensional spaces and are versatile in managing different types of data by leveraging various kernels, including linear, polynomial, and radial basis functions, etc. The equation for finding the hyperplane when the data is linearly separable is given by:

$$w_1x_1 + w_2x_2 + b = 0 \tag{5.1}$$

where, w_1 and w_2 are the weights corresponding to the features x_1 and x_2, and b is the bias term. The classifier predicts the label *y(x)* of a given input based on which side of the hyperplane it lies. The objective of the SVM is to maximize the margin between the two classes, which can be formulated as the following optimization problem [18]:

$$\begin{aligned} &\min_{w,b} \frac{1}{2}\|w\|^2 \\ &\text{Subject to the constraint}: y_i(w_1x_{i1} + w_2x_{i2} + b) \geq 1 \end{aligned} \tag{5.2}$$

Here, y_i represents the class label of the ith data point, and $\|w\|$ denotes the magnitude of the weight vector w, which controls the margin width.

When the data is not linearly separable, a kernel function $K(x_i, x_j)$ is employed to map the data into a higher-dimensional feature space.

The QSVM classifier is the quantum enhanced version of the classical SVM. By leveraging quantum feature spaces, QSVM can improve classification performance in certain complex datasets, offering advantages over classical SVM. In QSVM, classical data such as features related to climate science (e.g., total surface precipitation and surface soil wetness) are converted into quantum states through a quantum feature mapping method. The quantum feature map is expressed as [18]:

$$|\Phi(x)\rangle = \Upsilon_\theta(x)|0\rangle \tag{5.3}$$

where, $\Upsilon_\theta(x)$ is a unitary operator or quantum circuit that encodes the classical data x into the quantum state $\Phi(x)$, and $|0\rangle$ is the initial quantum state.

Similarly, for non-linear data points, in the case of QSVM, the kernel function measures the similarity between two data points using quantum mechanics. The quantum kernel is defined as [19]:

$$K_q(x_i, x_j) = |\langle\Phi(x_i)|\Phi(x_j)\rangle|^2 \tag{5.4}$$

Here, $\Phi(x_i)$ and $\Phi(x_j)$ represent the quantum states corresponding to the classical data points x_i, and x_j respectively; and the expression $\langle\Phi(x_i)|\Phi(x_j)\rangle$ denotes the inner product of these quantum states, whichreflects their similarity within the quantum feature space. The result of $|\langle\Phi(x_i)|\Phi(x_j)\rangle|^2$ determines how similar the two data points are in the quantum space.

QSVM is mostly used for tasks like identifying weather patterns or classifying climate regions. QSVM includes techniques and methods for studying climate science;

The quantum kernel support vector machine (QKSVM) converts the classical kernel function in SVMinto quantum states using a quantum kernel circuit thatsupports quantum effects to potentially improve accuracy [20–22].

The variational quantum SVM (VQSVM) uses variational quantum circuits (VQCs) to parameterize the kernel function and optimize the SVM decision boundary that allows for adaptive learning in high-dimensional spaces [23, 24].

The hybrid quantum–classical SVM is used to enhance the efficiency by integrating the classical SVM models with quantum optimization techniques or quantum kernel functions wherethe quantum part handles feature extraction, and the classical SVM handles the final classification [25, 26]. Particularly for large datasets thequantum annealing SVM (QA-SVM) uses quantum annealing to solve the SVM optimization problem and also can find the optimal hyperplanewithfaster and more precise than traditional methods [27]. To enhance the SVM's ability to distinguish non-linear patterns in climate data, the quantum feature map SVMuses quantum feature maps for embedding thetraditional data into quantum states [28, 29].

The application of QSVM in climate science is climate pattern classification (large climate datasets classified in different climate zones or weather types like cyclones, droughts, and heatwaves), extreme weather event prediction, anomaly detection in climate data (detecting unusual climate behavior such as sudden temperature changes or unexpected atmospheric pressure shifts), land use and vegetation classification (classifying satellite image to study the pattern of land use changes, vegetation cover, and deforestation over time) [30, 31], oceanic phenomenon analysis (identifying and predicting large-scale oceanic phenomena such as El Niño and La Niña), and renewable energy forecasting (classifies weather patterns to predict solar and wind energy availability) [32]. The QSVM can handle high-dimensional data better by using quantum kernels that map data into larger feature spaces, allowing more accurate separation of non-linear patterns, quickly processing massive datasets, detecting non-linear relationships, detecting anomalies in high-dimensional climate datasets, detecting bare soil, forecasting energy consumption, processing multispectral Earth observation data, offering accuracy and speed of satellite image classification by using quantum-enhanced feature extraction, and also process complex oceanic data with non-linear dependencies more effectively than classical SVMs [33].

5.3.2 Quantum Neural Networks

Quantum Neural Network (QNN) is a boon for climate science applications because they can process large-scale, high-dimensional datasets and model complex, non-linear relationships.

5.3.2.1 Quantum Convolutional Neural Networks (QCNNs)

The tasks like image classification and recognition, the QCNNs have been explored [34]. It uses quantum convolution layers to extract data features and the quantum pooling layers that are used to reduce dimensions from images which makes it effective for analyzing spatial data like satellite images [35].

QCNNs can classify satellite images for land cover, deforestation detection, and ice sheet monitoring, helping track environmental changes, processing satellite data to detect abnormal ocean temperature patterns, and aiding in early warnings for events like coral bleaching and El Niño. Figure 5.2 depicts a QCNN that applies multiple convolutional (CL) and pooling layers (PL) in sequence, ending with a fully connected layer (FCL) that performs a unitary transformation on the remaining qubits before measurement [36]. Here ρ_{in} and ρ_{out} represent the input and output quantum states, respectively.

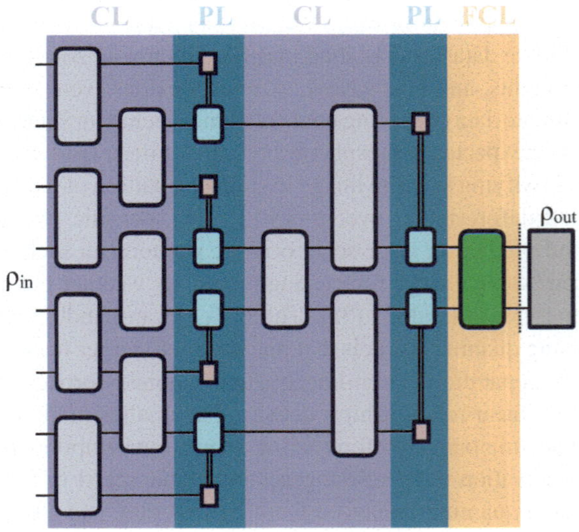

Fig. 5.2 A quantum convolutional neural network

5.3.2.2 Quantum Long Short-Term Memory (QLSTM)

In climate science, the complex time-series data is influenced by many interdependent environmental factors. To predict temperature, rainfall, and air quality traditional models like Recurrent Neural Networks (RNNs) and Long Short-Term Memory (LSTM) networks are mostly used. As data becomes more complex, these models need more computational power and often struggle with non-linear patterns. Quantum LSTM (Q-LSTM) improves classical LSTM by using quantum computing. This helps to handle complex and non-linear patterns more efficiently and predict with better accuracy.

Q-LSTM includes quantum kernelLSTM (QK-LSTM) whichenhances classical LSTM models by incorporating quantum kernel methods, allowing it to map input data into high-dimensional quantum feature spaces and this transformation helps in identifying complex, non-linear patterns in time-series data more effectively [34, 35]. Hybrid Quantum–Classical LSTMis a hybrid approachthat integrates quantum layers with classical LSTM layers, wherethe quantum part handles high-dimensional feature extraction, and the classical LSTM manages sequential data processing [37].

The QLSTMhelps in climate science by predicting extreme weather events, real-time monitoring, efficient complex system modeling, solar irradiance prediction, air quality forecasting, long-term climate modeling, faster processing of real-time climate data, ocean temperature prediction, and current prediction [38, 39].

Figure 5.3 shows the architecture of a QLSTM network, where data flows through multiple QLSTM layers. Each unit maintains a cell state (c) and hidden state (h) to process sequential informationefficiently. The input layer receives data at time t, and the processed information is passed through multiple QLSTM blocks before reaching the output layer which generates the final prediction.

5.3.2.3 Variational Quantum Neural Networks (VQNNs)

In VQNNs, the variational quantum circuits (VQCs) are integrated with classical neural networks. The VQCs are parameterized and trained using classical optimizers [40]. QCNNs can be applied in climate science to predict air quality, climate-driven extreme weather events, carbon emission reduction, renewable energy forecasting, and climate data analysis.

5.3.2.4 Quanvolutional Neural Networks (Quanvolutional Layers)

To extract important patterns more effectively, the Quanvolutional layers use random quantum circuits to change input data into quantum features. This is suitable or easier foranalyzing complex data, like satellite images, by capturing detailed spatial information. The application in climate science of quanvolutional neural networks can be weather pattern recognition, remote sensing image classification, forest fire detection, and ocean surface temperature analysis [41].

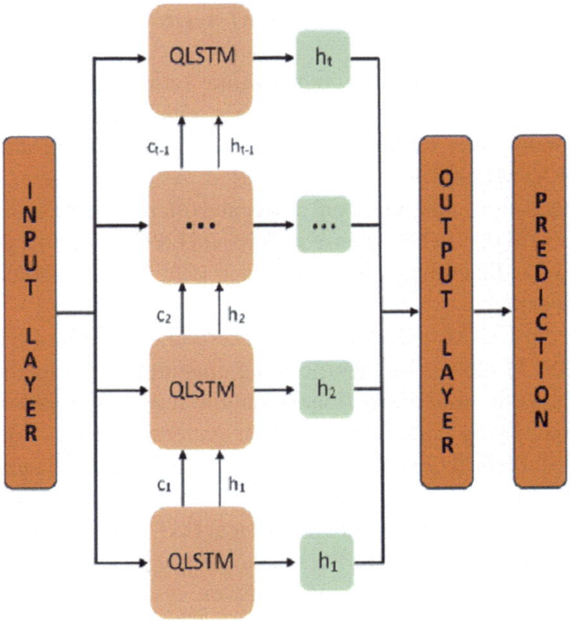

Fig. 5.3 Quantum long short-term memory model

5.3.2.5 Quantum Generative Adversarial Networks (QGANs)

Quantum Generative Adversarial Networks (QGANs) are the combination of GANs and quantum computing for improving generative learning. These are composed of two key components, quantum generator (a quantum model that creates new data samples by utilizing the unique properties of quantum mechanics) and classical or quantum discriminator (a model that maybe classical or quantum which evaluates the generated data and distinguishes it from real data) that leading to the creation of highly realistic data samples. A version called Wasserstein GAN (WGAN) uses Wasserstein distance for better training stability. The quantum version, QWGAN, is designed for current quantum computers, making quantum model training smoother and more scalable [10, 42]. The applications in climate science can be climate data simulation, and extreme event simulation.

5.3.2.6 Hybrid Quantum–Classical Neural Networks

These models combine classical neural networks with quantum layers. Classical layers handle initial data preprocessing, while quantum layers enhance feature extraction and prediction [43]. The application in climate science can be biodiversity impact assessment, carbon emission prediction.

5.3.3 *Quantum Principal Component Analysis (QPCA)*

The QPCA includes data preparation, covariance matrix construction, densitymatrix exponentiation, quantum phase estimation, eigenvalue and eigenvector extraction, principal component identification, and output reconstruction. Applications of QPCA in climate science can be climate modeling, real-time analysis, resource optimization, enhanced forecasting that can provide more accurate and detailed insights into climate patterns, and can process huge amounts of climate data more proficiently than the traditional methods [44].

5.3.4 *Harrow, Hassidim, and Lloyd (HHL)*

The HHL algorithm aims to solve linear equations represented in a quantum state. It is particularly useful in handling large datasets, performing machine learning tasks that involve linear equations, and offers significant efficiency improvements compared to classical methods [44]. Classical methods like Gaussian elimination solve linear systems with a complexity of $O(N^3)$, where N is the size of the system. HHL reduces this complexity to $O(\log(N))$ making it exponentially faster for large systems [45].

5.4 The Role of QML in Climate Science

QML can greatly contribute to climate science in many ways such as quickly analyzing large amounts of climate data, helping detect unusual patterns, helping us better understand how different climate factors interact, and making better decisions to fight against climate change. Managing renewable energy becomes easier with improved predictions, allowing better use of resources and reducing waste. They also help assess the impact of rising temperatures on agriculture, biodiversity, and coastal cities, providing useful insights for climate adaptation. Complex climate equations can be solved faster, improving large-scale climate models. Monitoring carbon emissions becomes more accurate, and industries can get recommendations for saving energy and reducing pollution.

Figure 5.4 presents a block diagram showing the components of the quantum weather forecast framework and their sequence. The way each block works, including the quantum circuits and algorithms used, may change depending on new scientific discoveries and specific needs. This framework helps us to understand that quantum computing can make weather predictions more precise, and reliable for climate changes and extreme weather events. This begins with data pre-processing, where raw weather data is cleaned and organized. Then, quantum feature extraction and quantum data processing prepare the data for quantum models. QSVM and QNN help analyze patterns and improve forecasting accuracy. Quantum clustering

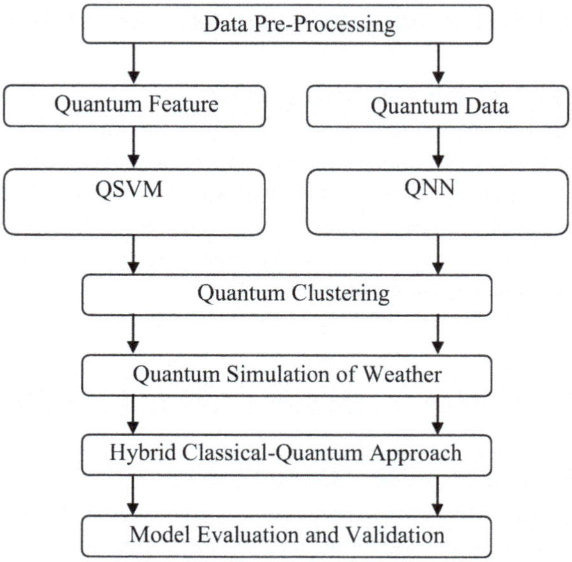

Fig. 5.4 Components of the quantum weather forecast framework

groups similar weather patterns, making it easier to predict regional climate changes. In the quantum simulation stage, quantum algorithms solve complex weather equations faster than traditional methods. Since quantum computers are still developing, a hybrid classical-quantum approach is used, combining classical weather models with quantum methods for better results. Finally, the model evaluation and validation stage checks the accuracy of forecasts and makes improvements.

QML technologies can also provide real-time tracking of extreme weather events like cyclones, droughts, hurricanes, and floods, leading to better early warning systems or models such as the weather research and forecasting (WRF) model where quantum optimization techniques can improve weather and climate models by making parameter uncertainty analysis more efficient, better simulations of regional climate patterns, and enhanced model performance across different climates and locations. Using algorithms like thequantum approximate optimization algorithm (QAOA), this is designed to tackle approximate solutions to complex optimization problems, especially any polynomial unconstrained binary optimization (PUBO) tasks [46]. Consider the following PUBO problem [47]:

$$\max_{y \in \{0,1\}^n} g(y) \tag{5.5}$$

With $g(y) : \{0, 1\}^{\mathrm{n}} \rightarrow \mathrm{R}$ and $\mathrm{g(y)} := \sum_{\mathrm{k}=1}^{\mathrm{m}} \mathrm{v}_k \mathrm{h}_k(\mathrm{y})$. Here, $v_k \in R$ and $m = O(poly(n))$ and Boolean functions $h_k(y)$ have a bounded degree such that $\deg(g) = d$. The QAOA provides approximate solutions for solving combinatorial optimization challenges.

QAOA is based on the principles of quantum adiabatic algorithm (QAA), which is designed to identifythe lowest eigenvalue, also known as the ground state energy, of a Hermitian matrix. The QAA process starts with a Hermitian matrix with a known minimum eigenstate and gradually evolves into another matrix with an unknown optimal solution state, tracking the transition throughout. However, QAA requires exponential time, making it computationally intensive. Moreover, its success probability does not always improve with longer execution times. In contrast, QAOA becomes more effective with each iteration, called a level, when the optimal parameters are applied.

QAOA begins with a quantum state $|r\rangle$, which is a superposition of all possible bit strings. A bit string $y$corresponds to a vector in the binary basis state $|y\rangle \in C^{2^n}$ where one entry is 1 while all others are 0. This superposition is created by applying Hadamard gates (H) to a zero-state $|0\rangle^{\otimes n}$ which is the tensor product of n zero qubits; that is

$$|r\rangle H^{\otimes n}|0\rangle^{\otimes n} = \frac{1}{\sqrt{2^n}} \sum_{y \in \{0,1\}^n} |y\rangle \tag{5.6}$$

Here, there are 2^n possible states where n represents the number of quantum bits. A QAOA circuit generates a bit string representing a solution. $F_p\ (\overrightarrow{\theta}, \overrightarrow{\alpha})$ is computed as the expected objective value from multiple runs of the algorithm. The number of levels influences the performance of QAOA with the optimal parameters, the algorithm asymptotically converges to the best solution as the number of levels increases.

Let $g^*: = max_y g(y)$ represent the optimal objective value and $F_p\ (\overrightarrow{\theta}, \overrightarrow{\alpha})$ be the expected output of the $\mathrm{QAOA_p}$. Then,

$$\lim_{p \to \infty} \max_{\vec{\theta}, \vec{\alpha}} F_p\left(\vec{\theta}, \vec{\alpha}\right) = g* \tag{5.7}$$

These tools also support disaster planning by optimizing resource distribution and minimizing losses. Decisionmakers can use data to create better plans for cutting emissions and protecting vulnerable areas. As technology advances, collaboration between climate experts and quantum scientists will lead to even better solutions and progress in climate research.

5.5 Conclusion and Future Work

In this chapter, we explored how quantum machine learning techniques and algorithms can help tackle climate change by identifying risks, planning for disasters, and cutting down carbon emissions. By improving quantum-based climate models and linking them with existing systems, we can make smarter, data-driven decisions to fight climate change. QML has the potential to be applied in real-world climate solutions and play a big role in creating a sustainable future. By improving quantum-based climate models and combining them with current systems, we can make better, data-driven decisions to fight climate change. Though QML holds immense potential, it is still in its early stages within the field of climate science. It faces challenges such as limited quantum computing power, the need for advanced algorithms, and the integration of QML with classical computing systems. In the future, we aim to tackle these challenges by refining QML techniques, developing practical algorithms, and improving the integration of quantum and classical systems for better results. Partnerships among quantum researchers, climate experts, and policymakers will be essential to develop effective solutions for combating climate change.

References

1. Singh, M., et al.: Quantum artificial intelligence for the science of climate change. In: Artificial Intelligence, Machine Learning and Blockchain in Quantum Satellite, Drone and Network, pp. 199–207. CRC Press (2022)
2. Parker, W.S.: Climate science. Elements Philos. Sci. (2024)
3. Dietz, T., Shwom, R.L., Whitley, C.T.: Climate change and society. Ann. Rev. Sociol. **46**(1), 135–158 (2020)
4. Rolnick, D., et al.: Tackling climate change with machine learning. ACM Comput. Surv. (CSUR) **55**(2), 1–96 (2022)
5. Choi, J., Oh, S., Kim, J.: The useful quantum computing techniques for artificial intelligence engineers. In: 2020 International Conference on Information Networking (ICOIN). IEEE (2020)
6. Beer, K., et al.: Training deep quantum neural networks. Nat. Commun. **11**(1), 808 (2020)
7. Biamonte, J., et al.: Quantum machine learning. Nature **549**(7671), 195–202 (2017)
8. Kong, F., Liu, X.-Y., Henao, R.: Quantum tensor network in machine learning: An application to tiny object classification (2021). arXiv:2101.03154
9. Schuld, M., Petruccione, F.: Supervised Learning with Quantum Computers, vol. 17. Springer, Berlin (2018)
10. Ramezani, S.B., et al.: Machine learning algorithms in quantum computing: a survey. In: 2020 International Joint Conference on Neural Networks (IJCNN). IEEE (2020)
11. Steane, A.: Quantum computing. Rep. Prog. Phys. **61**(2), 117 (1998)
12. Rietsche, R., et al.: Quantum computing. Electron. Markets **32**(4), 2525–2536 (2022)
13. Abdelgaber, N., Nikolopoulos, C.: Overview on quantum computing and its applications in artificial intelligence. In: 2020 IEEE Third International Conference on Artificial Intelligence and Knowledge Engineering (AIKE). IEEE (2020)
14. Kumar, S., Simran, S., Singh, M.: Quantum intelligence: merging AI and quantum computing for unprecedented power. In: 2024 International Conference on Trends in Quantum Computing and Emerging Business Technologies. IEEE (2024)
15. Peral-García, D., Cruz-Benito, J., José García-Peñalvo, F.: Systematic literature review: quantum machine learning and its applications. Comput. Sci. Rev. **51**, 100619 (2024)
16. Nammouchi, A., Kassler, A., Theorachis, A.: Quantum Machine Learning in Climate Change and Sustainability: a Review (2023). arXiv:2310.09162
17. Suthaharan, S., Suthaharan, S.: Support vector machine. In: Machine Learning Models and Algorithms for Big Data Classification: Thinking with Examples for Effective Learning, pp. 207–235 (2016)
18. Munasinghe, T., et al.: Assessment of quantum ML applicability for climate actions: comparison of the variational quantum classifier and the quantum support vector classifier with classical ML models. In: 2024 IEEE International Conference on Big Data (BigData). IEEE (2024)

19. Havlíček, V., et al.: Supervised learning with quantum-enhanced feature spaces. Nature **567**(7747), 209–212 (2019)
20. Sridevi, S., et al.: Quantum enhanced support vector machine with instantaneous quantum polynomial encoding for improved cyclone classification. In: 2023 6th International Conference on Recent Trends in Advance Computing (ICRTAC). IEEE (2023)
21. Miroszewski, A., et al.: Detecting clouds in multispectral satellite images using quantum-kernel support vector machines. IEEE J. Sel. Top. Appl. Earth Observations Remote Sens. (2023)
22. Zhou, X., et al.: Quantum kernel estimation-based quantum support vector regression. Quantum Inf. Process. **23**(1), 29 (2024)
23. Innan, N., et al.: Enhancing quantum support vector machines through variational kernel training. Quantum Inf. Process. **22**(10), 374 (2023)
24. Yi, J., et al.: Variational Quantum Linear Solver enhanced Quantum Support Vector Machine (2023). arXiv:2309.07770
25. Orazi, F., et al.: Hybrid quantum technologies for quantum support vector machines. Information **15**(2), 72 (2024)
26. Masum, A.K.M., et al.: Hybrid quantum-classical machine learning for sentiment analysis. In: 2023 International Conference on Machine Learning and Applications (ICMLA). IEEE (2023)
27. Delilbasic, A., et al.: A single-step multiclass SVM based on quantum annealing for remote sensing data classification. IEEE J. Sel. Top. Appl. Earth Observations Remote Sens. (2023)
28. Chen, B.-S., Chern, J.-L.: Generating quantum feature maps for SVM classifier (2022). arXiv: 2207.11449
29. Suhas, S., Divya, S.: Quantum-improved weather forecasting: integrating quantum machine learning for precise prediction and disaster mitigation. In: 2023 International Conference on Quantum Technologies, Communications, Computing, Hardware and Embedded Systems Security (iQ-CCHESS). IEEE (2023)
30. Wijata, A.M., et al.: Detection of bare soil in hyperspectral images using quantum-kernel support vector machines. In: IGARSS 2024–2024 IEEE International Geoscience and Remote Sensing Symposium. IEEE (2024)
31. Gupta, M.K., Romaszewski, M., Gawron, P.: Potential of quantum machine learning for processing multispectral Earth observation data. Authorea Preprints (2023)
32. Rezaei, T., Javadi, A.: Environmental impact assessment of ocean energy converters using quantum machine learning. J. Environ. Manag. **362**, 121275 (2024)
33. Nutakki, M., Koduru, S., Mandava, S.: Quantum support vector machine for forecasting house energy consumption: a comparative study with deep learning models. J. Cloud Comput. **13**(1), 1–12 (2024)
34. Fan, F., et al.: Hybrid quantum-classical convolutional neural network model for image classification. IEEE Trans. Neural Netw. Learn. Syst. (2023)
35. Lin, C.-H., Lin, T.-H., Chanussot, J.: Quantum information-empowered graph neural network for hyperspectral change detection. IEEE Trans. Geosci. Remote Sens. (2024)
36. Mangini, S., et al.: Quantum computing models for artificial neural networks. Europhys. Lett. **134**(1), 10002 (2021)
37. Hsu, Y.-C., Li, T.-Y., Chen, K.-C.: Quantum Kernel-Based Long Short-term Memory (2024). arXiv:2411.13225
38. Lin, C.-H.A., Liu, C.-Y., Chen, K.-C.: Quantum-train long short-term memory: application on flood prediction problem. arXiv:2407.08617
39. Chen, S.Y.-C., Yoo, S., Fang, Y-L.L.: Quantum long short-term memory. In: ICASSP 2022–2022 IEEE International Conference on Acoustics, Speech and Signal Processing (ICASSP). IEEE (2022)
40. Das, M., Bolisetti, T.: Variational quantum neural networks (VQNNS) in image classification (2023). arXiv:2303.05860
41. Sebastianelli, A., et al.: Quanv4EO: Empowering Earth Observation by Means of Quanvolutional Neural Networks (2024). arXiv:2407.17108
42. Tsang, S.L., et al.: Hybrid quantum-classical generative adversarial network for high resolution image generation. IEEE Trans. Quantum Eng. (2023)

43. Dong, Y., et al.: Air quality prediction based on quantum activation function optimized hybrid quantum classical neural network. Front. Phys. **12**, 1412664 (2024)
44. Rahman, S.M., et al.: Climate change through quantum lens: computing and machine learning. Earth Syst. Environ. 1–18 (2024)
45. Cho, C.-H., et al.: Quantum computation: Algorithms and applications. Chin. J. Phys. **72**, 248–269 (2021)
46. Choi, J., Kim, J.: A tutorial on quantum approximate optimization algorithm (QAOA): Fundamentals and applications. In: 2019 International Conference on Information and Communication Technology Convergence (ICTC). IEEE (2019)
47. Fakhimi, R., Validi, H.: Quantum Approximate Optimization Algorithm (QAOA). In: Encyclopedia of Optimization, pp. 1–7. Springer International Publishing, Cham (2023)

Chapter 6
Quantum Cloud Computing: Key Technologies, Challenges, and Opportunities

Anisha Kumari, Ranjan Kumar Behera, and Bibhudatta Sahoo

Abstract Quantum cloud computing represents a groundbreaking fusion of quantum computing and cloud technology, unlocking new possibilities for advanced computation and data processing. This work explores the current landscape of the field, highlighting key developments and ongoing challenges. We examine major advancements, including novel quantum algorithms, hybrid computing models, and improvements in quantum cloud infrastructure. Additionally, critical challenges such as security concerns, resource allocation complexities, and the integration of quantum technologies into existing cloud architectures are addressed. A comparative analysis of Infrastructure as a Service (IaaS), Platform as a Service (PaaS), and Quantum as a Service (QaaS) is also presented, offering insights into their distinct roles and capabilities. The discussion further extends to major quantum cloud platforms, such as IBM Quantum Experience and Amazon Braket, and their contributions to innovation and accessibility. Moreover, we explore the transformative impact of quantum cloud computing across industries, including cryptography, material science, and artificial intelligence. This chapter provides valuable insights for researchers and industry professionals, enabling future advancements in quantum cloud computing.

Keywords Cloud computing · Quantum cloud · Qubits · Quantum gates · Cloud architecture · Quantum-as-a-service

---

A. Kumari
Department of CSE, KIIT University, Bhubaneswar, Odisha, India
e-mail: anishamishracs@gmail.com

R. K. Behera (✉)
Department of CSE, IIIT Ranchi, Ranchi, Jharkhand, India
e-mail: ranjan@iiitranchi.ac.in

B. Sahoo
Department of CSE, NIT Rourkela, Rourkela, Odisha, India
e-mail: bdsahu@nitrkl.ac.in

S. Dehuri et al. (eds.), *Advances in Quantum Inspired Artificial Intelligence*, Intelligent Systems Reference Library 274, https://doi.org/10.1007/978-3-031-89905-8_6

6.1 Introduction

Quantum computing, based on the principles of quantum mechanics, signifies a significant advancement in computational capability. Quantum computers differ from classical computers by utilizing qubits, which can exist in superposition states, allowing for the simultaneous representation of multiple values, rather than processing information solely in binary states (0s and 1s) [1, 2]. This unique capability enables quantum systems to enhance the performance at an exponential rate as compared to classical systems in various applications, including optimization, factorization, and the simulation of quantum systems.

Conversely, cloud computing has revolutionized the accessibility of computational resources by enabling on-demand delivery of services such as storage, processing power, and platforms via the internet. The cloud paradigm provides scalability, cost efficiency, and accessibility, democratizing advanced computational tools for users worldwide. The coupling of quantum computing with the technology of clouds commonly referred to as Quantum Cloud Computing (QCC) [3]. Quantum cloud platforms enable researchers, developers, and enterprises to experiment with quantum methodologies and solve domain-specific problems without requiring physical access to quantum hardware, which remains costly and highly specialized [4]. Companies such as IBM, Google, Microsoft, and Amazon have spearheaded this integration by offering platforms like IBM Quantum Experience, Google's Quantum AI, Amazon Braket, and Microsoft Azure Quantum, which facilitate quantum experimentation and innovation through the cloud [5].

Quantum cloud computing is more than a technological curiosity; it serves as a critical enabler for addressing computational challenges in various domains including cryptography, medical science, financial modeling, and artificial intelligence [6, 7]. By combining quantum computing's unparalleled processing potential with the accessibility of cloud services, QCC significantly lowers entry barriers for organizations and researchers, fostering innovation in areas previously constrained by classical computing capabilities. Despite its immense promise, QCC is still in its infancy. The field grapples with significant challenges, including hardware limitations such as qubit coherence times, error rates, and scalability, as well as cloud-specific concerns like data security, latency, and efficient resource allocation [8, 9]. Understanding these challenges and analyzing ongoing advancements are essential to unlocking the complete capabilities of such transformative technology. With the rapid pace of development, a systematic review is necessary to synthesize knowledge across disciplines, identify emerging trends, and highlight critical gaps. This chapter seeks to present a synthesis of state-of-the-art advancements in QCC, an analysis of key challenges hindering the field's progress, and guidance on future research directions and innovation opportunities. By addressing these objectives, this chapter contributes as a comprehensive resource for researchers, technologists, and policymakers, offering insights into the current state of QCC and the pathways for its advancement.

The primary contributions of the chapter are outlined as follows.:

- The significance of quantum computing within the cloud environment is explored, emphasizing its role in addressing complex computational problems.
- Key issues and challenges in quantum cloud computing, including hardware limitations and cloud-specific concerns, are thoroughly presented.
- Major tools and platforms offering quantum processing capabilities in the cloud, such as IBM Quantum Experience and Amazon Braket, are discussed in detail.
- A comparative analysis among PaaS, IaaS abd QaaS is presented to indicate the advancement in QaaS.
- Future directions for research and development are proposed, focusing on innovation strategies, industry adoption, and policy recommendations.

This chapter is structured in the following manner. In Sect. 6.2, the basics of quantum and cloud computing is presented in order to understand the integration process. Section 6.3 introduces the architectural blueprint for quantum cloud computing. Then, in Sect. 6.4, how quantum computing blends with different cloud technologies has been explored. Section 6.5 highlights some of the popular frameworks and tools that make quantum cloud computing possible. Moving forward, Sect. 6.6 discuss the wide variety of real-world applications this technology enables. Section 6.7 discusses exciting research opportunities and potential directions for future exploration. Finally, Sect. 6.8 wraps up the chapter with a summary and the future perspective in this evolving field.

6.2 Fundamental of Quantum Computing

Quantum computing signifies a fundamental transformation from classical computing by harnessing fundamental quantum mechanical principles like entanglement, superposition, and quantum interference. These phenomena enable quantum computers to process information in ways that far surpass classical systems, making them particularly suited for solving complex problems that are infeasible for traditional computing architectures.

Superposition enables quantum bits (qubits) to occupy multiple states at the same time rather than being confined to binary states of 0 or 1 [10]. This property exponentially increases the parallelism of computations, enabling quantum computers to perform multiple calculations at once. Entanglement is another key feature, where qubits become interdependent regardless of physical distance [11]. This correlation allows for synchronized operations that significantly enhance computational efficiency, enabling applications such as secure quantum communication and powerful optimization algorithms. Quantum interference further refines computational processes by amplifying correct computational paths while diminishing erroneous ones [12]. This property is crucial in algorithms like Grover's search algorithm, where interference helps identify optimal solutions faster than classical counterparts.

These quantum phenomena empower quantum systems to tackle highly complex problems, including large-scale optimizations, cryptographic analysis, and molecular simulations. Some of the terminology used in quantum mechanics are described as below.

6.2.1 Qubits

A qubit, or quantum bit, serves as the basic form of information in quantum computing, paralleling the role of a classical bit in traditional computing. Qubits possess distinct characteristics as a result of quantum mechanics principles. The key features of Qubits are defined as below:

- Superposition: A qubit differs from a classical bit in that it can occupy a state of superposition, rather than being limited to a binary value of 0 or 1. This means it can be in a combination of both 0 and 1 simultaneously. In simple terms, we can describe the state of a qubit using this equation:

$$|\psi\rangle = \alpha|0\rangle + \beta|1\rangle \tag{6.1}$$

  Here, α and β are complex numbers that tell us the likelihood of the qubit being measured as either 0 or 1. These can be considered as weights that determine the probabilities of each outcome.
- Entanglement: Qubits exhibit entanglement, indicating that the state of one qubit may be contingent upon the state of another, regardless of the spatial separation between them. This allows for the creation of correlations that classical bits cannot achieve, enabling powerful computing capabilities.
- Inference: Quantum algorithms often use interference to amplify the probabilities of correct results and diminish the probabilities of incorrect ones. This property enables the efficient solution of certain complex problems.
- Measurement: Upon measurement, a qubit transitions from its superposition state to one of the basis states (0 or 1), with the probabilities dictated by the amplitudes α and β. The mentioned measuring process is fundamentally different from classical measurement.

6.2.2 Quantum Gates

Quantum gates are the basic tools that work with qubits in quantum circuits, doing things that regular computer gates can't. While classical gates change bits in a one-way, irreversible manner, quantum gates work differently-they use reversible transformations that keep the overall probabilities of quantum states intact. This unique behavior is rooted in the rules of quantum mechanics. By harnessing quantum phenomena like superposition and entanglement, these gates enable calculations that go far beyond what classical computers can achieve. Some of the most commonly used quantum gates include:

- Pauli Gates (X, Y, Z) [13]:These fundamental gates serve as quantum counterparts to classical procedures, incorporating further phase alterations.
- The Hadamard gate [14] is crucial for quantum computing as it creates superposition, allowing qubits to exist in a linear combination of $|1\rangle$ and $|0\rangle$. Hadamard gates serve a crucial function in quantum parallelism, allowing multiple computational paths to be explored simultaneously in algorithms like Shor's and Grover's.
- CNOT Gate [15]: The Controlled-NOT (CNOT) gate entangles two qubits, a crucial property for quantum computation. This gate is essential for teleportation and error correction.
- Toffoli [16]: A three-qubit gate where the third qubit is flipped if the first two qubits are $|1\rangle$. This gate is significant in quantum error correction and is universal for classical reversible computing.
- Fredkin Gates [17]: It Swaps two qubits based on the state of a control qubit. It ensures logical reversibility and is useful in quantum algorithms that require conditional swaps.

6.2.3 Quantum Algorithms

Quantum algorithms harness the fundamental principles of superposition, entanglement, and quantum interference to tackle problems far beyond the reach of classical computing. By leveraging quantum parallelism, these algorithms enable computations that would be impractical or even impossible for traditional systems. Below, we explore some of the most significant quantum algorithms and their real-world applications.

- Shor's Algorithm [18]: Introduced by Peter Shor in 1994, this groundbreaking algorithm revolutionized cryptography by demonstrating how quantum computing can efficiently factor large integers. Unlike classical methods, which require exponential time-approximately $O(2^n)$. Shor's algorithm achieves this in polynomial time, with a complexity of $O(n^3)$. This efficiency is made possible through the use of quantum Fourier transforms to identify periodic patterns in modular exponentiation, a capability that poses a serious challenge to widely used encryption systems like RSA.
- Grover's Algorithm [19]: One of the popular Quantum search algorithm was proposed by Grover in 1996 to reduce the time complexity for seraching an element in unsorted databases. This algorithm can be used to solve complex optimization problem. The time complexity for searching an element in a database having n

number of unsorted element is reduced to $O(\sqrt{N})$ as compared to best classical algorithm which is around $O(N)$.

- The Quantum Approximate Optimization Algorithm (QAOA) [20] was created especially to solve combinatorial optimization issues, which are important in fields including operations research, finance, logistics, and machine learning. These issues entail selecting the optimal option from a limited number of options, frequently while dealing with intricate limitations.

  As a hybrid quantum-classical algorithm, QAOA iteratively improves solutions by fusing quantum computation with traditional optimization methods. It makes use of parameterized quantum circuits, in which a series of quantum gates that are reliant on adjustable parameters evolve a quantum state. These parameters are then adjusted using a classical optimizer to increase the likelihood of measuring an ideal or nearly ideal solution.

6.3 Architectural Model of Quantum Cloud Compuing

The Quantum Cloud Computing Framework is a sophisticated system that integrates classical and quantum computing resources to provide scalable, accessible, and reliable quantum computing services. The framework for quantum cloud computing is shown in Fig. 6.1. The detailed explanation of each component in the framework is presented as below:

1. **User Interface**: The User Interface is the entry point for users to interact with the quantum cloud computing system. It includes web portals, command-line interfaces (CLIs), and integrated development environments (IDEs). These interfaces allow users to submit quantum tasks, such as quantum circuits or algorithms, and retrieve results. The user interface is designed to be intuitive and user-friendly, enabling both experts and non-experts to leverage quantum models. Developers can access pre-built library for quantum processing for optimization, machine learning, and other applications, making it easier to solve complex problems without deep quantum expertise.
2. **Quantum Cloud Service Layer**: The Quantum Cloud Service Layer acts as the bridge between users and the underlying quantum computing infrastructure. It provides APIs, SDKs, and libraries for quantum algorithm development [21]. Popular quantum programming frameworks like Qiskit (IBM) [22], Cirq (Google) [23], and PennyLane (Xanadu) [24] are integrated into this layer. The service layer validates and processes user-submitted tasks, ensuring they are compatible with the available quantum hardware. It also offers pre-built quantum algorithms and tools for users to develop custom solutions.
3. **Orchestration and Management Layer**: The Orchestration and Management Layer is responsible for managing the execution of quantum tasks. It schedules tasks based on the availability of quantum and classical resources, ensuring efficient utilization of the system. This layer allocates classical compute resources

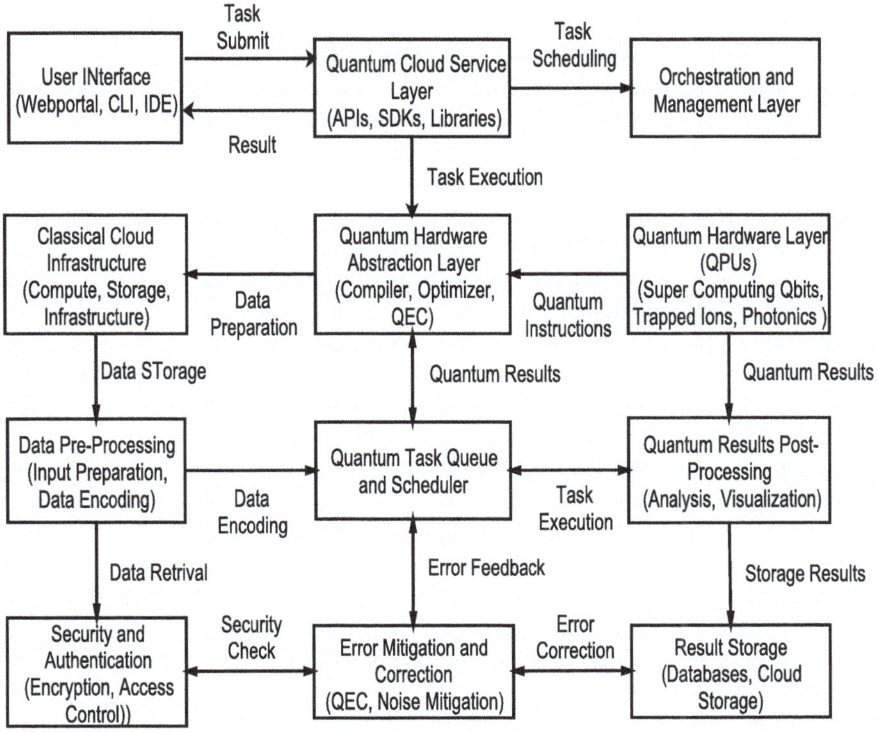

Fig. 6.1 Quantum cloud architectural model

(e.g., CPUs, GPUs) for pre-processing and post-processing tasks, while also managing access to quantum hardware (e.g., QPUs). Additionally, it implements error mitigation and correction strategies to improve the reliability of quantum computations. The orchestration layer plays a key role in optimizing hybrid quantum-classical workflows, ensuring seamless integration between classical and quantum components.

4. **Classical Cloud Infrastructure**: The Classical Cloud Infrastructure provides the backbone for the quantum cloud computing system. It includes compute resources (e.g., CPUs, GPUs), storage systems (e.g., databases, object storage), and networking infrastructure. This layer supports pre-processing tasks, such as data preparation and encoding, as well as post-processing tasks, such as result analysis and visualization. The classical infrastructure ensures scalability, reliability, and security, enabling the system to process complicated workflow by handling big data systems. It also facilitates low-latency communication between classical and quantum components, ensuring efficient task execution.
5. **Quantum Hardware Abstraction Layer**: The Quantum Hardware Abstraction Layer translates high-level quantum algorithms into low-level instructions that can be executed on quantum hardware. It includes a quantum compiler, which

converts quantum circuits into quantum assembly language (QASM) [25], and a circuit optimizer, which optimizes circuits for specific hardware architectures. This layer also implements quantum error correction (QEC) [26] techniques to mitigate errors caused by noise and decoherence. By abstracting the complexities of quantum hardware, this layer enables users to run quantum tasks without needing deep knowledge of the underlying hardware.

6. **Quantum Hardware Layer**: The Quantum Hardware Layer consists of physical quantum processors (QPUs) that execute quantum tasks [7]. These processors are based on various technologies, such as superconducting qubits (e.g., IBM, Google), trapped ion qubits (e.g., IonQ), and photonic qubits (e.g., Xanadu). The quantum hardware executes the low-level instructions generated by the abstraction layer and produces results. Due to the inherent noise and instability of quantum systems, this layer often requires cryogenic systems (for superconducting qubits) and precise control electronics to maintain qubit coherence.
7. **Data Pre-processing**: The Data Pre-Processing component prepares input data for quantum tasks. This includes encoding classical data into quantum-compatible formats, such as quantum states or amplitudes. Pre-processing may also involve data normalization, feature extraction, or other transformations to ensure the data is suitable for quantum algorithms. This component interacts closely with the classical cloud infrastructure to retrieve and store data efficiently. Effective pre-processing is essential for guaranteeing the precision as well as effectiveness of quantum calculations.
8. **Quantum Task Queue and Scheduler**: The Quantum Task Queue and Scheduler manages the queue of quantum tasks submitted by users. It schedules tasks for execution based on the availability of quantum resources and prioritizes tasks according to predefined criteria. This component interacts with the orchestration layer to allocate resources and optimize task execution. It also handles error feedback from the error mitigation component, ensuring that tasks are retried or corrected as needed. The scheduler plays a crucial role in maintaining the efficiency and reliability of the quantum cloud computing system.
9. **Quantum Result Post-processing**: The Quantum Result Post-Processing component analyzes and interprets the results returned by the quantum hardware. This may involve statistical analysis, machine learning, or other techniques to extract meaningful insights from the raw quantum data. The results are then visualized in a user-friendly format, such as graphs or charts, to help users understand and interpret the outcomes. Post-processed results are stored in the result storage component for future retrieval and analysis.
10. **Result Storage**: The Result Storage component stores the results of quantum computations in scalable databases or cloud storage systems. This allows users to retrieve and analyze results at any time, even after the quantum task has been completed. The storage system is designed to handle large volumes of data and ensure data integrity and security. It interacts closely with the classical cloud infrastructure to provide seamless access to stored results.

6.4 Quantum Computing and Cloud Integration

Cloud computing enables instant access to computing power, storage, and other resources via the internet, allowing users to scale and utilize services as needed without relying on local infrastructure. This model has been transformative for industries by democratizing access to cutting-edge technologies, enabling rapid scalability, and fostering innovation in sectors such as healthcare, finance, education, and research. For instance, cloud computing facilitates collaboration by allowing global teams to access shared platforms, supports cost-effective solutions by replacing capital expenditures with operational expenditures, and accelerates digital transformation by integrating advanced tools like artificial intelligence and big data analytics. These impacts extend far beyond its technological backbone, shaping how organizations operate and compete in a rapidly evolving digital landscape. This paradigm is crucial for quantum computing, as it mitigates accessibility and scalability barriers.

6.4.1 Cloud Architecture

Cloud computing is transforming organizational operations by offering scalable, adaptable, and economical alternatives. Fundamentally, cloud computing is built on a multi-layered architecture that ensures seamless integration of resources, platforms, and applications [27, 28]. These layers enable efficient management, development, and usage of cloud-based services, catering to diverse needs across industries. The architecture comprises three primary layers: the infrastructure layer, the platform layer, and the application layer. Each layer plays a critical role in delivering the overall functionality of cloud computing. The different layered in cloud architecture is shown in Fig. 6.2.

6.4.1.1 Infrastructure Layer

The infrastructure layer forms the foundation of cloud computing. Often referred to as Infrastructure as a Service (IaaS), it provides the virtualized hardware resources necessary for cloud operations [29]. Key components of this layer include:

- Virtualized Computing Resources: The infrastructure layer offers virtual machines, CPUs, GPUs, and memory to handle diverse computational workloads. Virtualization technology ensures that physical resources are optimally utilized and dynamically allocated based on demand.
- Storage: This layer provides scalable storage solutions, including block storage, object storage, and file systems. These storage solutions are essential for managing data-intensive operations and ensuring high availability.
- Networking: Advanced networking capabilities, such as virtual private networks (VPNs), load balancers, and software-defined networking (SDN), are integral to

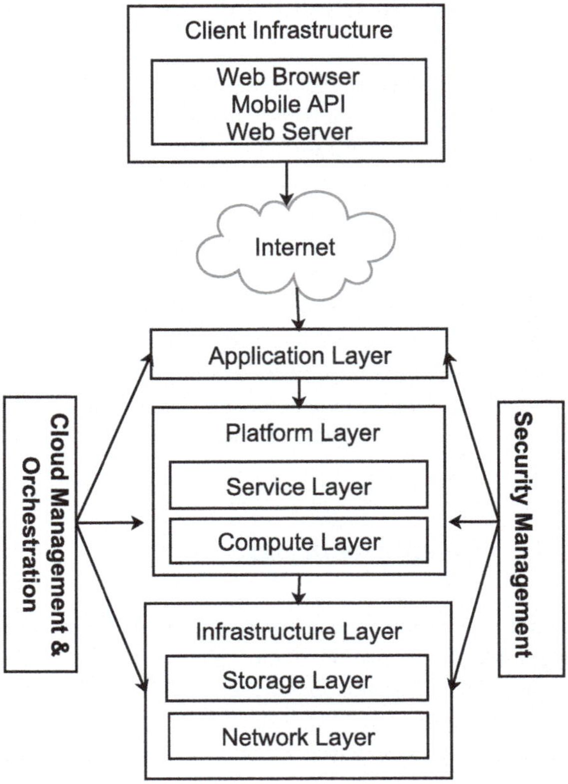

Fig. 6.2 Cloud architecture

this layer. They ensure secure and efficient communication within and across cloud environments.

The infrastructure layer allows organizations to bypass the complexities of managing physical hardware, reducing operational costs and enabling rapid scalability. Examples of IaaS providers include Amazon Web Services (AWS) EC2, Google Compute Engine, and Microsoft Azure Virtual Machines [30].

6.4.1.2 Platform Layer

The platform layer, often referred to as Platform as a Service (PaaS), sits above the infrastructure layer and provides environments for application development, testing, and deployment. This layer is particularly valuable for developers, offering tools and frameworks that simplify the software development lifecycle [31].

- Development Tools: The platform layer includes integrated development environments (IDEs), code editors, and libraries to accelerate the creation of applications. These tools are often pre-configured, eliminating the need for extensive setup.

- Application Hosting: PaaS solutions provide runtime environments for hosting applications, ensuring compatibility and performance optimization.
- Database Management: Managed databases, such as relational and NoSQL databases, are accessible through the platform layer. These services include automatic scaling, backups, and maintenance.

PaaS solutions empower organizations to streamline development processes, reduce time-to-market, and foster innovation. Examples of PaaS offerings include Google App Engine, Microsoft Azure App Services, and Salesforce's Heroku.

6.4.1.3 Application Layer

The application layer, also known as Software as a Service (SaaS), represents the topmost layer of the cloud computing architecture. It hosts software services and applications which are directly accessible to the clients through web browsers or mobile applications.

- User-Friendly Interfaces: SaaS applications are designed with intuitive interfaces, enabling users to perform tasks without technical expertise. These applications cater to a wide range of needs, from productivity tools to enterprise software.
- Accessibility: Being cloud-based, SaaS solutions are accessible from anywhere with an internet connection. This ensures mobility and flexibility for users.
- Subscription Models: Most SaaS providers operate on subscription-based pricing, offering users cost-effective access to powerful software without requiring upfront investments.

6.4.2 Integration of Quantum Computing with Cloud Platforms

The integration of quantum computing with cloud platforms is a transformative development, bridging cutting-edge quantum technology with the accessibility and scalability of cloud computing. This convergence allows users to access quantum resources alongside classical computational tools, opening new avenues for addressing complex problems in science, engineering, and industry. By offering seamless interfaces and managed services, cloud platforms make quantum computing practical and accessible to a broad audience, including researchers, developers, and businesses, regardless of their expertise in quantum mechanics or the need for specialized hardware management.

Cloud platforms achieve this integration through key elements such as hybrid computing and abstracted hardware management. The hybrid computing model enables users to utilizing software development kits (SDKs) to simplify quantum programming and simulate quantum behavior on classical infrastructure. At the same time, cloud integration abstracts the complexities of quantum hardware, allowing users to interact with quantum processors via intuitive interfaces or APIs. This abstraction empowers users to focus on developing quantum algorithms and applications without the burden of managing the intricate operational requirements of quantum systems.

Several platforms are at the forefront of this integration, providing a diverse range of tools, services, and hardware access:

- IBM Quantum Experience: It is a leading cloud platform that lets users explore and experiment with IBM's quantum processors, making quantum computing more accessible for research and development. These range from small-scale systems to cutting-edge devices with over 100 qubits, enabling researchers and developers to explore the frontiers of quantum technology. A cornerstone of the platform is the Qiskit open-source SDK, which allows users to design, simulate, and execute quantum circuits using Python. With Qiskit, developers can build and test quantum algorithms across a variety of applications, fostering a practical understanding of quantum mechanics and computation.
- Microsoft Azure Quantum: Microsoft Azure Quantum integrates quantum computing capabilities seamlessly into the Azure cloud ecosystem, offering tools for hybrid quantum-classical workflows. It supports multiple quantum hardware backends, including ion trap systems and superconducting qubits, as well as quantum-inspired algorithms designed for classical systems. The platform's Quantum Development Kit (QDK) provides developers with a robust environment to program in high-level languages such as Q# and Python, enabling them to tackle optimization problems in domains like financial modeling, supply chain management, and energy grid optimization. By integrating these quantum capabilities with Azure's broader cloud services, Microsoft Azure Quantum empowers businesses to incorporate quantum technology into their existing infrastructure and solve real-world challenges more efficiently.
- Amazon Braket: Amazon Braket is a managed quantum computing service that supports the entire lifecycle of quantum algorithm development, from prototyping to execution. It provides accessibility to diverse quantum hardware platforms, like gate-based quantum computers, quantum annealers, and simulators, catering to a wide range of research and application needs. Amazon Braket also includes tools for circuit visualization and debugging, simplifying the process of testing and refining quantum algorithms. This flexibility allows users to explore quantum computing at their own pace, advancing their understanding of the technology and its potential applications.

The integration of quantum computing with cloud platforms profoundly impacts industries by addressing shortcomings beyond the capabilities of classical computing. In pharmaceuticals, quantum simulations of molecular configurations and chemical reactions accelerate pharmaceutical development. In finance, quantum computing optimizes investment portfolios, improves risk management, and enhances fraud detection. Logistics benefits from quantum solutions for large-scale optimization problems such as route planning and supply chain management, while the energy sector uses quantum tools to optimize renewable energy grids and advance battery material discovery. Nevertheless, attaining the entire potential of this integration requires surmounting several obstacles. Scalable cloud infrastructures must be built to seamlessly handle quantum and classical resources, and standardized protocols and APIs are needed for consistent access to quantum systems across platforms. Quantum hardware limitations, such as noise and error rates, demand advanced error mitigation techniques, while a skilled workforce equipped with quantum expertise is essential to harness this technology effectively.

6.4.3 Quantum Cloud Service Models

The emergence of cloud-based quantum computing has revolutionized the availability of quantum resources, enabling individuals and organizations to leverage this cutting-edge technology without needing direct ownership or management of quantum hardware. Three distinct service models-Infrastructure as a Service (IaaS), Platform as a Service (PaaS), and Quantum as a Service (QaaS)-play critical roles in this ecosystem [32]. Each model provides a unique level of abstraction and caters to specific user needs, from researchers requiring direct hardware access to developers seeking simplified environments for quantum application development. Below, we explore each model in detail and their relevance to quantum computing. A comparative analysis of IaaS, PaaS and QaaS is presented in Table 6.1.

6.4.3.1 Infrastructure as a Service (IaaS) for Quantum Computing

Infrastructure as a Service (IaaS) grants users direct access to quantum hardware through cloud-based APIs [33]. This model is ideal for researchers and developers who need to experiment with real quantum devices to test and refine their quantum algorithms. Leading examples of IaaS in quantum computing include IBM Quantum and Rigetti Quantum Cloud Services [34], which provide users with access to superconducting qubit-based quantum processors. The following steps can be used to integrate the IaaS with quantum services:

- Users interact directly with quantum hardware through cloud interfaces, often submitting quantum circuits for execution via APIs.
- Cloud providers handle the complex backend, including cryogenic environments and error correction systems required to maintain the quantum processors.
- Researchers benefit from the ability to run experiments on real quantum hardware, observe quantum phenomena, and compare results with simulated outputs.

IaaS offers significant flexibility and control over quantum hardware, making it ideal for several use cases. It is particularly advantageous for advanced quantum algorithm research, where studying the physical behavior of qubits on real hardware is essential for gaining insights into quantum mechanics and computational techniques.

Table 6.1 A comparative analysis between IaaS, PaaS and QaaS

Feature	IaaS	PaaS	QaaS
Objctives	Provides virtualized computing resources over the cloud	Delivers a platform featuring tools for application development	Provides remote access to quantum computing capabilities through cloud-based platforms
Computing power	Uses classical computing hardware (CPUs, GPUs)	Uses classical computing with optimized development tools	Leverages quantum computing power (qubits, quantum circuits)
Performance	Limited by Moore's Law and classical processing speeds	Faster development but still bound by classical processing limits	Exponentially faster for certain problem domains (e.g., cryptography, optimization, simulations)
Use cases	General-purpose computing, hosting applications, storage, and networking	Application development, Data analytics, Machine learning frameworks	Quantum simulations, Cryptographic analysis, Drug discovery, AI optimizations
Complexity	Requires management of virtual machines, Networking, and Storage	Simplifies development but still relies on classical architectures	Highly complex but abstracts quantum hardware for user-friendly access
Scalability	Scalable, but increases in demand require more traditional computing resources	Scalable for applications but limited by classical computing efficiency	Quantum parallelism enables superior scalability for solving large-scale computational problems
Security	Relies on encryption, VPNs, and Access control	Secured at the application and platform levels but vulnerable to classical cyber threats	Quantum cryptography can provide unbreakable encryption (e.g., Quantum Key Distribution—QKD)
Cost efficiency	Pay-as-you-go model but high costs for high-performance needs	Reduces development costs but still expensive for large applications	High initial cost, but long-term efficiency due to quantum speedups
Technology dependency	Requires traditional server-based computing infrastructure	Relies on a mix of cloud services and traditional computing	Uses Quantum Processors (QPU) integrated into cloud services, requiring specialized hardware
Adoption readiness	Fully mature and widely adopted	Well-adopted with extensive development support	Emerging but rapidly evolving; Expected to dominate future high-performance computing

Additionally, IaaS facilitates the testing of hardware-specific optimization strategies, such as refining error correction protocols or improving gate fidelity, which are critical for advancing quantum computing technology. Moreover, IaaS serves an important role in educational contexts, providing students and academic researchers with hands-on access to actual quantum systems, enabling them to deepen their understanding of quantum computing by experimenting with real-world hardware rather than relying solely on simulations.

6.4.3.2 Platform as a Service (PaaS) for Quantum Computing

It offers pre-configured quantum programming environments, software development kits (SDKs), and simulators [35]. Developers can focus on writing and testing quantum algorithms without worrying about the underlying hardware. Notable examples include Google Cirq, Microsoft Azure Quantum, and Qiskit from IBM. The working principal of Paas in Quantum computing is as below:

- Users access quantum programming tools through cloud-based platforms, which often include SDKs and IDEs tailored for quantum development.
- These tools come with quantum simulators that allow developers to test algorithms on classical infrastructure before deploying them on quantum hardware.
- PaaS platforms may also offer pre-built quantum algorithms and libraries, simplifying the development of common use cases like optimization and cryptography.

It reduces development time by providing ready-to-use libraries, pre-built examples, and development tools that streamline the creation of quantum applications. PaaS platforms let developers test algorithms in simulations before deployment, reducing errors. They also enable hybrid workflows by integrating quantum and classical computing for more efficient problem-solving.

6.4.3.3 Quantum as a Service (QaaS)

Quantum as a Service (QaaS) takes abstraction to the highest level, allowing users to execute quantum algorithms without requiring detailed knowledge of quantum programming or hardware [36]. This model is particularly beneficial for businesses and researchers who want to leverage quantum computing for specific applications without delving into the complexities of quantum mechanics. Prominent examples of QaaS platforms include Amazon Braket and D-Wave Leap. The working principle for Quantum as a service service model can be as below:

- Users define their problem in a high-level interface, such as submitting an optimization problem or a quantum circuit.
- The QaaS platform abstracts the complexity of translating the problem into a quantum representation, managing hardware execution, and returning the results.

- These services often include algorithm templates, such as those for optimization, machine learning, or cryptography, making quantum computing accessible even to non-specialists.

QaaS democratizes access to quantum computing by removing the steep learning curve traditionally associated with the field, making it accessible to a wider audience, including those with little to no quantum expertise. It is particularly well-suited for industry applications such as optimizing supply chains, financial modeling, and material discovery, where quantum computing can provide significant advantages over classical approaches. Additionally, QaaS supports rapid prototyping of quantum algorithms with minimal development effort, allowing organizations to experiment and innovate without requiring deep technical knowledge of quantum systems. This makes QaaS an attractive option for businesses and institutions exploring quantum computing capabilities without the need to invest heavily in specialized talent or infrastructure.

6.5 Frameworks and Architectures Enabling Quantum Cloud Computing (QCC)

Integrating quantum computing into cloud systems is crucial for making quantum technology more accessible, enabling researchers and industry professionals to leverage its capabilities for advanced problem-solving. Cloud platforms have emerged as the gateway to quantum computing, providing accessible, scalable, and cost-effective solutions for those looking to harness the power of quantum processors without the need to own or manage the complex hardware. Several frameworks and platforms are key to facilitating this integration, enabling the design, simulation, and execution of quantum algorithms.

6.5.1 Quantum SDKs: Bridging Classical and Quantum Computing

Quantum SDKs are fundamental to the hybridization of cloud with quantum systems, as they provide essential software tools for developing quantum algorithms, simulating quantum circuits, and interfacing with quantum hardware. These SDKs allow developers and researchers to write, debug, and test quantum programs on classical machines before executing them on actual quantum processors. Some notable quantum SDKs include Qiskit, PyQuil, and Cirq.

- Qiskit: Developed by IBM, Qiskit is an open-source SDK that provides tools for quantum computing and quantum programming. is used to develop classical and quantum simulators, hardware integration with IBM's quantum processors,

and tools for quantum algorithm optimization. Researchers use Qiskit to perform quantum error correction, study quantum systems, and build quantum applications across various industries. Qiskit supports integration with IBM Quantum Experience, providing cloud access to real quantum processors.
- PyQuil: Developed by Rigetti Computing, PyQuil is another Python-based SDK that supports quantum circuit design, quantum algorithm development, and interaction with quantum hardware. PyQuil provides access to Rigetti's quantum computing platform, Forest, which includes simulators and real quantum processors. PyQuil allows for hybrid quantum-classical computations, enabling researchers to design models which can execute in both on traditional computers and quantum processors.
- Cirq: Google's Cirq is an open-source quantum computing framework specifically designed for quantum circuit design, simulation, and optimization, with a focus on near-term quantum devices. Cirq provides tools for developing quantum algorithms that can be executed on Google's quantum processors, including those used in Google Quantum AI. It also integrates well with other cloud computing services, allowing for hybrid quantum-classical applications.

These SDKs offer crucial abstractions that allow quantum programmers to interact with quantum processors without needing to understand the deep underlying physics of quantum mechanics. They simplify the development process, streamline algorithm optimization, and provide tools for testing quantum circuits (Table 6.2).

6.5.2 Hybrid Quantum-Classical Architectures

Quantum hardware is a major obstacle in quantum computing. Current quantum computers fall into the noisy intermediate-scale quantum (NISQ) regime, which has a finite number of qubits that can be manipulated but suffers from high levels of error. Hybrid quantum-classical architectures have been developed to help address these limitations. They leverage the strengths of classical systems as well as quantum ones and distribute the computational tasks according to the unique capabilities of the two categories of systems.

In a hybrid quantum-classical architecture, quantum processors handle the most computationally intensive parts of a problem-typically those involving complex quantum states or large-scale optimization-while classical processors handle the rest of the computation. This division allows users to leverage the power of quantum computing for certain tasks while relying on the reliability and scalability of classical computing for other parts of the algorithm.

For instance, the quantum computer may utilize to accelerate parts of the learning process, such as generating quantum feature maps or solving optimization problems, while the classical computer can handle data preprocessing, model training,

Table 6.2 A comparative analysis of frameworks used in cloud quantum computing

Tool/Simulator/Framework	Primary function	Provider	Key features	Cloud access	Programming language	Use cases
IBM quantum experience	Quantum computing platform	IBM	Access to real quantum processors, Qiskit SDK, Quantum simulations	Yes	Python (Qiskit)	Quantum algorithm development, Research, Education
Google cirq	Quantum circuit development	Google	Optimized for near-term quantum algorithms, Tight hardware integration	No direct cloud, Via Google cloud AI	Python	Quantum algorithm prototyping, Hardware-specific research
Amazon braket	Cloud-based quantum computing	AWS	Multi-vendor access (Rigetti, IonQ, OQC), Hybrid computing	Yes	Python (Boto3, PennyLane)	Cloud-based quantum simulations, Algorithm testing
Microsoft azure quantum	Quantum cloud ecosystem	Microsoft	Supports multiple quantum providers, Q#, Quantum-inspired optimization	Yes	Q# (via QDK), Python	Optimization, Materials science, Quantum algorithm development
Rigetti forest	Hybrid quantum-classical computing	Rigetti	Quil language, Cloud-accessible quantum hardware	Yes	Python, Quil	Quantum circuit simulations, Hybrid computing
D-Wave Leap	Quantum annealing and optimization	D-Wave	Focus on quantum annealing, Hybrid solver service	Yes	Python (Ocean SDK)	Optimization problems, AI/ML, logistics
ProjectQ	Quantum compiler and simulation	ETH Zurich	Hardware-agnostic quantum compiler, Integrates with Qiskit	No (local simulator)	Python	Compilation of quantum circuits, Hybrid quantum-classical computing
PennyLane	Quantum Machine Learning (QML)	Xanadu	Focus on QML, Supports multiple backends	Yes (via AWS Braket, IBM, etc.)	Python	Quantum AI/ML, Variational quantum circuits
Qiskit aer	Quantum circuit simulation	IBM	High-performance quantum circuit simulation	No (local simulator)	Python	Testing quantum algorithms before deployment
QuEST	High-performance quantum simulation	University of Oxford	GPU-accelerated quantum simulation for large-scale systems	No	C/C++	Large-scale quantum simulations, Benchmarking

and result analysis. Such architectures are often supported by hybrid quantum-classical frameworks, which facilitate the smooth interaction between the two types of systems.

6.5.3 Quantum Simulators

Quantum simulators are software tools that allow developers to emulate quantum circuits on classical hardware. These simulators are crucial for debugging, testing, and optimizing quantum algorithms without the need to run them on actual quantum hardware. They provide a cost-effective way to experiment with quantum circuits before executing them on real quantum devices, which are often limited in terms of accessibility, speed, and cost.

For example, IBM Qiskit Aer is a simulator provided by IBM Quantum that allows users to simulate quantum circuits on classical computers. Qiskit Aer offers both state vector simulators and noise models, enabling developers to test quantum algorithms under ideal and noisy conditions. This is important because quantum hardware often experiences noise, and simulating noisy environments allows developers to test and refine error correction techniques. Similarly, PyQuil offers the Forest simulator, which provides tools for simulating quantum programs and testing algorithms in a controlled environment before running them on Rigetti's quantum processors.

Simulators are essential for the development process, enabling developers to fine-tune their algorithms, troubleshoot issues, and refine their approaches before they commit to running them on actual quantum hardware. They also facilitate the scaling of quantum algorithms, as they allow testing on larger quantum circuits than what may be feasible with current hardware.

6.6 Applications of Quantum Cloud Computing

Cloud quantum computing represents a paradigm shift in computational power, allowing researchers, developers, and industries to use capabilities of models without the need for specialized hardware. It provide scalable, accessible, and cost-effective solutions for utilizing quantum resources. The ability to harness quantum computing via the cloud opens up a range of applications across multiple fields, from chemistry and physics to finance and logistics.

6.6.1 Pharmaceutical and Healthcare Advancements

Quantum computing is poised to revolutionize the pharmaceutical industry by accelerating pharmaceutical discoveries, optimizing molecular modeling, and enabling enhanced accurateness in the simulation of intricate biological systems. Traditionally, the development of novel pharmaceuticals involves time-consuming or expensive processes, with researchers relying on classical simulations to model molecular interactions. However, classical computers often struggle with accurately simulating the quantum effects that govern molecular behavior, leading to approximations that limit the potential of drug discovery.

Quantum devices possess the capability to directly model quantum states, potentially significantly decreasing the time required for simulating the structures of molecules as well as responses. Quantum algorithms enable academics to simulate the actions of large molecules with exceptional accuracy, thereby enhancing the efficiency of the drug discovery process. By leveraging quantum resources via cloud platforms, pharmaceutical companies can perform large-scale molecular simulations, identify potential drug candidates, and optimize drug design.

In healthcare, the interactions between proteins and other molecules in the human body can be mimic by quantum computers, helping to develop personalized treatments for various diseases. Quantum-based algorithms may be employed for genomics, facilitating increased effective data analysis, pattern recognition, and disease prediction models.

6.6.2 Financial Services and Risk Analysis

In the financial sector, quantum computing offers significant potential to improve optimization, risk analysis, and portfolio management. Quantum computers are particularly suited to solving optimization problems, such as finding the best portfolio allocation or the most efficient asset distribution across multiple investments. These problems often involve large datasets and require solving highly complex equations, which classical computers struggle to do efficiently as the number of variables increases.

Quantum Annealing and the Quantum approximation algorithms are particularly useful for portfolio optimization, helping investors identify the best combination of investments while minimizing risk and maximizing returns. Additionally, it may financial modeling and derivatives pricing by simulating the behavior of financial markets with greater accuracy. These simulations can be used to model market volatility, optimize investment strategies, and improve fraud detection by analyzing large datasets in ways that classical algorithms cannot match.

By leveraging quantum resources available through cloud platforms, financial institutions can accelerate the analysis of vast amounts of financial data, reducing the time needed to make critical investment decisions and improving risk management strategies.

6.6.3 Machine Learning and Artificial Intelligence

Over the past decade, machine learning (ML) and artificial intelligence (AI) have made remarkable progress, finding their way into countless areas of our lives. From understanding and generating human language to enabling computers to "see" and interpret images, and even powering self-driving cars, these technologies are transforming how we live and work. However, classical computing approaches often face limitations when performing complex calculations or processing large datasets in a reasonable time frame. It can enhance certain stages in machine learning processes, including pattern recognition, data clustering, and optimization. Quantum-enhanced machine learning methods can markedly decrease the time needed for model training and enhance their accuracy. Quantum algorithms can enhance the performance of neural networks and decision trees, leading to accelerated and more efficient AI models. Moreover, quantum computing may facilitate novel methodologies. for solving problems that classical computers struggle with, such as large-scale unsupervised learning tasks and the training of complex models involving high-dimensional data. Cloud-based quantum computing platforms can facilitate the deployment as well as the development of quantum-augmented machine learning algorithms, enabling companies to unlock the potential of quantum AI without needing to invest heavily in quantum hardware.

6.6.4 Cryptography and Cybersecurity

The topic of cryptography is one of the most well-known and extensively researched uses of quantum computing. RSA and ECC (Elliptic Curve Cryptography), two popular encryption techniques today, are based on mathematical problems that are computationally challenging for traditional computers to solve. However, by solving these issues at an exponentially faster rate than traditional computers, quantum computers-using methods like Shor's Algorithm-may be able to crack various encryption techniques. While cybersecurity is seriously threatened by quantum computing, it also opens the door for creative developments in quantum-resistant encryption. It makes it easier to develop stronger security protocols, such as post-quantum cryptography (PQC) and quantum key distribution (QKD). These methods ensure defence against prospective quantum-based cyberthreats by utilising the basic ideas of quantum physics to create extremely secure communication channels.

In order to prepare for a future in which quantum computers could be able to crack current encryption algorithms, cloud quantum computing platforms can help with the development and deployment of quantum-safe cryptography systems. Utilising cloud-based quantum resources allows companies to stay ahead of the curve and safeguard their data from possible quantum threats.

6.6.5 Optimization in Logistics and Supply Chain Management

Optimization is a critical challenge in industries such as logistics, transportation, and supply chain management. Classical optimization techniques often struggle to find the best solution when dealing with complex, high-dimensional problems, such as determining the most efficient route for delivery trucks, managing inventory, or optimizing manufacturing processes. Quantum computing, with its ability to handle large, complex datasets and explore multiple solutions simultaneously. For example, quantum computers could help logistics companies optimize delivery routes, minimizing fuel consumption and reducing costs. Similarly, quantum computing could be applied to optimize SCM in improving inventory management, forecasting, and production scheduling. Cloud platforms offering quantum computing resources make these powerful optimization tools accessible to businesses of all sizes, allowing them to enhance operational efficiency and gain a competitive edge in their industries.

6.6.6 Climate Modeling and Environmental Science

Quantum computing holds promise for addressing complex environmental challenges, particularly in the areas of climate modeling and resource management. Climate models are notoriously difficult to simulate due to the vast number of variables involved, such as atmospheric conditions, ocean currents, and weather patterns. Classical computing struggles to simulate these models at the necessary level of detail, often relying on approximations that may not fully capture the underlying physical processes.

Quantum computing could enable more accurate climate simulations by directly modeling the quantum effects that govern weather patterns and climate systems. Quantum algorithms could simulate the behavior of particles and molecules in the atmosphere, providing more detailed and accurate predictions of climate change. This could help researchers better understand the impacts of global warming, improve weather forecasting, and optimize the use of natural resources.

Moreover, quantum computing can be used to develope energy efficient model and explore new ways to capture and store renewable energy. By leveraging cloud-based quantum computing resources, environmental scientists can accelerate their research and contribute to efforts to combat climate change.

6.7 Research opportunities

Quantum cloud computing is a cutting-edge area that blends the power of quantum computing with the flexibility of cloud technology, allowing users to access quantum tools and resources from anywhere, anytime. By providing quantum processors as cloud-based services, organizations and researchers can leverage quantum computing without requiring direct access to expensive and complex hardware. However, several challenges remain, including scalability, security, performance optimization, and interoperability with classical cloud infrastructures. Addressing these challenges presents numerous research directions and opportunities that can help shape the developing models in Quantum cloud systems. Some of research directions are presented as below:

- **Scalable Quantum Cloud Infrastructure**: Developing scalable infrastructure that integrates quantum and classical computing resources efficiently could be one future research direction. Since quantum processors (qubits) are highly sensitive and require controlled environments, managing and distributing quantum workloads over the cloud remains a challenge. Researchers are exploring hybrid architectures where classical and quantum computing systems work together, dynamically assigning tasks to the most suitable computing resource. Additionally, optimizing quantum hardware resource allocation in cloud environments can improve execution efficiency, minimize costs, and enhance user accessibility.
- **Quantum Cloud Security and Privacy**: Security and privacy are major challenges in quantum cloud computing. This is because quantum computers could one day crack the encryption methods we currently rely on. To tackle this, researchers are working on creating stronger, quantum-resistant encryption techniques to protect traditional cloud systems from quantum-based attacks. At the same time, quantum technologies like quantum key distribution (QKD) are showing great potential for enabling ultra-secure communication in quantum clouds. Beyond that, scientists are exploring ways to use quantum protocols for secure collaboration and developing privacy-focused quantum algorithms to keep data safe in cloud-based quantum systems.
- **Quantum Software-as-a-Service (QSaaS)**: To make quantum computing accessible to a broader audience, research is focused on developing Quantum Software-as-a-Service (QSaaS) platforms. These platforms provide cloud-based quantum programming environments, APIs, and toolkits that allow users to develop, test, and deploy quantum algorithms without requiring expertise in quantum hardware. Standardizing quantum software development kits (SDKs), improving compiler efficiency, and creating algorithm repositories for cloud quantum computing can further accelerate the adoption of quantum computing across industries.
- **Quantum Resource Virtualization**: Similar to classical cloud computing, virtualization is an essential component of quantum cloud services. Quantum resource virtualization aims to enable dynamic resource sharing across multiple quantum processors and users. Researchers are working on virtualized quantum machines that allow for more efficient execution of quantum workloads, reducing latency and

improving quantum cloud scalability. Additionally, noise mitigation and quantum error correction techniques are essential for maintaining computational accuracy in cloud-based quantum computing, making this an important area of study.

- **Distributed Quantum Networking**: To unlock the true power of quantum cloud computing, we need to make significant progress in two key areas: quantum networking and distributed quantum computing. These advancements will help connect quantum systems and enable them to work together seamlessly, paving the way for more powerful and efficient quantum solutions. Research in this area focuses on integrating quantum communication protocols, such as quantum teleportation and entanglement-based networking, into cloud environments. The development of the quantum internet, which enables secure and efficient data transfer between quantum cloud nodes, is a promising direction. Distributed quantum computing, where multiple quantum computers work together across the cloud, could enhance computational power and reliability, making it possible to tackle more complex problems.
- **Performance Optimization and Benchmarking**: As quantum cloud computing evolves, performance optimization and benchmarking become crucial for evaluating quantum cloud services. Researchers are working on defining metrics to assess execution speed, error rates, and resource efficiency in cloud-based quantum computing. Adaptive error correction strategies, dynamic qubit allocation, and resource-efficient compilation techniques are key areas of focus to improve quantum cloud service performance. By establishing standardized benchmarks, researchers can compare different quantum cloud platforms and enhance their reliability and efficiency.
- **Quantum-AI Hybrid Cloud Systems**: Combining quantum computing with artificial intelligence (AI) presents a significant research opportunity, particularly in cloud environments. Quantum computing has the potential to enhance AI applications by accelerating machine learning algorithms. Researchers are exploring how cloud-based quantum computing can improve AI-driven tasks such as deep learning, natural language processing, and predictive analytics.
- **Edge-Quantum Cloud Computing**: The performance of edge computing can be improved by integrating it with quantum computing. Edge-quantum cloud computing explores how quantum computing resources can be leveraged at the edge to enable faster and more secure data processing for real-time applications. Potential use cases include secure IoT networks, real-time financial modeling, and adaptive AI-driven systems. Research is needed to develop frameworks that effectively integrate quantum computing with edge devices and classical cloud infrastructure.

6.8 Conclusion and Future Works

Quantum cloud computing stands at the intersection of two transformative fields, promising unparalleled computational power and accessibility. This review has highlighted significant advancements in quantum hardware, software, and applications, with platforms such as IBM Quantum, Google Quantum AI, Amazon Braket, and Microsoft Azure Quantum leading the way. These platforms have made quantum computing accessible to a broader audience, enabling researchers and enterprises to experiment with cutting-edge technologies. However, challenges related to scalability, error rates, security, and economic feasibility continue to impede the widespread adoption of quantum cloud services. To overcome these hurdles, targeted investments in research and development are essential. Efforts should focus on building fault-tolerant quantum systems, optimizing resource allocation in multi-tenant environments, and enhancing the usability of quantum development tools. Additionally, fostering global collaboration among academia, industry, and governments will be critical for addressing resource inequities and driving innovation. From advancing scientific discovery to transforming industries, its impact will be far-reaching. By addressing current challenges and embracing collaboration, quantum cloud computing can unlock a future of unprecedented possibilities, propelling humanity into a new era of technological achievement.

References

1. Neumann, N.M.P., van der Schoot, W., Sijpesteijn, T.: Quantum Cloud Computing from a User Perspective, pp. 236–249. Springer, Berlin (2023)
2. Patel, T., et al.: Toward privacy in quantum program execution on untrusted quantum cloud computing machines for business-sensitive quantum needs (2023). arXiv:2307.16799
3. Golec, M., et al.: Quantum cloud computing: trends and challenges. J. Econ. Technol. (2024)
4. Rahaman, M., Islam, M.M.: A review on progress and problems of quantum computing as a service (qcaas) in the perspective of cloud computing. Global J. Comput. Sci. Technol. **15** (2015)
5. Castelvecchi, D.: Ibm's quantum cloud computer goes commercial. Nature **543** (2017)
6. Singh, H., Sachdev, A.: The Quantum Way of Cloud Computing, pp. 397–400. IEEE (2014)
7. Nguyen, H.T., Krishnan, P., Krishnaswamy, D., Usman, M., Buyya, R.: Quantum cloud computing: a review, open problems, and future directions (2024). arXiv:2404.11420
8. Soeparno, H., Perbangsa, A.S.: Cloud quantum computing concept and development: a systematic literature review. Procedia Comput. Sci. **179**, 944–954 (2021)
9. Grossi, M., et al.: A serverless cloud integration for quantum computing (2021). arXiv:2107.02007
10. Kovachy, T., et al.: Quantum superposition at the half-metre scale. Nature **528**, 530–533 (2015)
11. Fasel, S., Halder, M., Gisin, N., Zbinden, H.: Quantum superposition and entanglement of mesoscopic plasmons. New J. Phys. **8**, 13 (2006)
12. Kumari, A., Sahoo, B., Behera, R.K.: Mitigating Cold-start Delay Using Warm-start Containers in Serverless Platform, pp. 1–6. IEEE (2022)
13. Jena, A., Genin, S., Mosca, M.: Pauli partitioning with respect to gate sets (2019). arXiv:1907.07859

14. Shepherd, D.J.: On the role of hadamard gates in quantum circuits. Quantum Inf. Process. **5**, 161–177 (2006)
15. Bataille, M.: Quantum circuits of cnot gates: optimization and entanglement. Quantum Inf. Process. **21**, 269 (2022)
16. Monz, T., et al.: Realization of the quantum toffoli gate with trapped ions. Phys. Rev. Lett. **102**, 040501 (2009)
17. Patel, R.B., Ho, J., Ferreyrol, F., Ralph, T.C., Pryde, G.J.: A quantum fredkin gate. Sci. Adv. **2**, e1501531 (2016)
18. Ugwuishiwu, C., Orji, U., Ugwu, C., Asogwa, C.: An overview of quantum cryptography and shor's algorithm. Int. J. Adv. Trends Comput. Sci. Eng **9** (2020)
19. Gong, C., et al.: Grover algorithm-based quantum homomorphic encryption ciphertext retrieval scheme in quantum cloud computing. Quantum Inf. Process. **19**, 1–17 (2020)
20. Fakhimi, R., Validi, H.: Quantum Approximate Optimization Algorithm (qaoa), pp. 1–7. Springer, Berlin (2023)
21. Zeydan, E., Baranda, J., Mangues-Bafalluy, J.: Post-quantum blockchain-based secure service orchestration in multi-cloud networks. IEEE Access **10**, 129520–129530 (2022)
22. Cross, A.: The ibm q experience and qiskit open-source quantum computing software **2018**, L58–003 (2018)
23. Younis, E., Iancu, C.: Quantum Circuit Optimization and Transpilation Via Parameterized Circuit Instantiation, pp. 465–475. IEEE (2022)
24. Bergholm, V., et al.: Pennylane: Automatic differentiation of hybrid quantum-classical computations (2018). arXiv:1811.04968
25. Khammassi, N., et al.: cqasm v1. 0: Towards a common quantum assembly language (2018). arXiv:1805.09607
26. Aoki, T., et al.: Quantum error correction beyond qubits. Nat. Phys. **5**, 541–546 (2009)
27. Varia, J.: Cloud architectures. White Paper of Amazon, http://jineshvaria.s3.amazonaws.com/public/cloudarchitectures-varia.pdf **16** (2008)
28. Kumari, A., Sahoo, B., Behera, R.K.: Workflow aware analytical model to predict performance and cost of serverless execution. Concurr. Comput. Practice Exp. **35**, e7743 (2023)
29. Yildiz, M., Abawajy, J., Ercan, T., Bernoth, A.: A Layered Security Approach for Cloud Computing Infrastructure, pp. 763–767. IEEE (2009)
30. Innocent, A.: Cloud infrastructure service management-a review (2012). arXiv:1206.6016
31. Guoli, Z., Wanjun, L.: The Applied Research of Cloud Computing Platform Architecture in the E-Learning Area, vol. 3, pp. 356–359. IEEE (2010)
32. Kumari, A., Sahoo, B.: Serverless Architecture for Healthcare Management Systems, pp. 203–227. IGI Global (2022)
33. Bhardwaj, S., Jain, L., Jain, S.: Cloud computing: a study of infrastructure as a service (iaas). Int. J. Eng. Inf. Technol. **2**, 60–63 (2010)
34. Karalekas, P.J., et al.: A quantum-classical cloud platform optimized for variational hybrid algorithms. Quantum Sci. Technol. **5**, 024003 (2020)
35. Rani, D., Ranjan, R.K.: A comparative study of saas, paas and iaas in cloud computing. Int. J. Adv. Res. Comput. Sci. Softw. Eng. **4** (2014)
36. Garcia-Alonso, J., et al.: Quantum software as a service through a quantum api gateway. IEEE Internet Comput. **26**, 34–41 (2021)

Chapter 7
Security, Privacy, and Trust in Quantum Edge-Driven Intelligence: Challenges and Solutions

Sucheta Panda and Sushree Bibhuprada B. Priyadarshini

Abstract Quantum Computing (QC) along with Edge-Driven Intelligence (EDI) is becoming a powerful paradigm for decentralized decision-making and data processing driven by the explosive growth of Internet of Things (IoT) appliances and the massive growth of data generated at the network edge. However, there are also a lot of security, privacy, and trust-related issues that come with executing EDI. Some of the most pressing issues our world is facing could be resolved by quantum computing, include those related to materials science, energy, climate, agriculture, health, and the environment, etc. As the size of the system increases, classical computing becomes more difficult to solve some of these issues. Thus, considering the security, privacy, and trust in the context of QC and EDI, this abstract delves into the main problems and factors encountered in present era of technology. We address the particular risks to security such as data leaks, illegal access, and Denial-of-Service (DoS) attacks that are brought about by the distributed nature of edge computing environments. We have also discussed the privacy consequences of handling sensitive data at the edge including location and health data, and the importance of putting strong privacy-preserving measures in place. In order to create safe and reliable quantum-edge computing ecosystems, we have outlined the potential approaches and solutions for boosting security, protecting privacy, and building faith in the quickly changing field of edge-driven intelligence.

Keywords Quantum computing · Decentralization · Edge-driven intelligence · Data breaches · Quantum communication · IoT · Privacy · Security

---

S. Panda · S. B. B. Priyadarshini (✉)
Siksha O' Anusandhan Deemed to Be University, Bhubaneswar, Odisha, India
e-mail: sushreepriyadarshini@soa.ac.in

S. Dehuri et al. (eds.), *Advances in Quantum Inspired Artificial Intelligence*, Intelligent Systems Reference Library 274, https://doi.org/10.1007/978-3-031-89905-8_7

7.1 Introduction

In contrast to relying solely on centralized data centers or cloud servers, edge computing is a decentralized computing approach in which data processing and storage are carried out close to the data source or endpoint devices. According to this concept, computer power is situated at the "edge" of the network, usually inside or close to the data-generating or-consuming devices. Being nearer to the data source allows quicker processing, less latency, and more efficiency as compared to conventional cloud-based or centralized computing techniques. A few of the factors driving the emergence of edge computing are the rapidly increasing number of Internet of Things (IoT) devices along with the massive increase in data created at the network edge, and the growing need for real-time data processing and analysis. Edge computing offers several benefits, including improved scalability, reduced bandwidth limitations, and more effective use of network resources by dispersing computing resources over a network [1, 2].

The increasing usage of edge computing has ushered in an era of decentralized intelligence in the ever-changing field of modern technology. By putting computer resources close to the data source, this paradigm minimizes latency, maximizes scalability, and enables real-time processing. But as businesses use edge-driven intelligence to spur creativity and productivity, they also have to face up the enormous obstacles related to trust, security, and privacy. The convergence of processing, data, and communication at the edge creates a wide range of complicated vulnerabilities that jeopardize sensitive data's secrecy and integrity. Furthermore, because edge settings are dynamic and heterogeneous, putting strong security measures in place can be difficult and often call for creative solutions to protect infrastructure and data. In this connection, transformational improvements in the processing, storing, and use of data have resulted from the development of computing technologies. The capacity of classical edge computing to process data closer to its source, lowering latency, speeding up reaction times, and boosting data security have conjointly made it more popular. By utilizing quantum mechanical phenomena like superposition, entanglement, and interference, quantum computing has simultaneously become a disruptive force that can solve problems that were previously thought to be computationally impossible. The next development in this line of thinking is edge quantum computing. The goal of this paradigm is to increase computational capacity with decreasing value of latency for quantum-based jobs, and to offer a more secure and effective data-processing framework by combining quantum computing units with edge computing networks.

The applications of edge computing architectures are very diverse; they can be found in large-scale applications like autonomous vehicles, smart cities, and telecom networks, or small-scale ones in industrial sensors and smart home gadgets. Edge computing refers to a collection of technologies that allow applications to operate closer to the source of data generation or consumption. These technologies include edge servers, edge gateways, edge routers, and edge computing platforms [3–5]. A significant ethical and technological problem is to strike a balance between the demands of data usefulness and people's rights to control their personal information [6]. In addition, adherence to legal frameworks like the CCPA and GDPR adds to the complexity of the situation by demanding careful consideration of governance procedures and privacy-preserving measures. Building and preserving confidence in edge-driven intelligence ecosystems becomes essential in the face of these difficulties. The dependability, morality, and ethics of edge systems and the organizations that run them must be trusted by stakeholders. Furthermore, in order to maintain a culture of trust, active collaboration between end users, lawmakers, and industry stakeholders is becoming necessary for effective risk reduction and the handling of emerging threats [7]. Within this framework, this study investigates the complex aspects of privacy, security, and edge-driven intelligence trust aspect. We analyze the changing threat landscape, new best practices, and potential paths for research and innovation by incorporating information from regulatory agencies, industry, and academia. We aim to provide stakeholders with the knowledge and tools that they need to handle the complexities of edge computing in a way that is ethical, responsible, and safe by exploring these important concerns [8–10]. The operation of Edge computing is explained in Fig. 7.1. Additionally, new technologies like quantum communications, quantum sensing, and quantum computing may be able to close the performance gaps left by their classical counterparts.

The idea of a Quantum Edge Computing (*QEC*) simulator is a platform which is presented here for creating apps for the upcoming generation of edge computing. In order to enable research on quantum edge applications, a QEC simulator is intended to incorporate components from edge computing and quantum technology. The growing need for sensors that are more accurate and sensitive while using less power to provide bigger and denser datasets is what spurred the idea that was presented. The below equations serve as the cornerstone of quantum computing, illustrating the ideas and

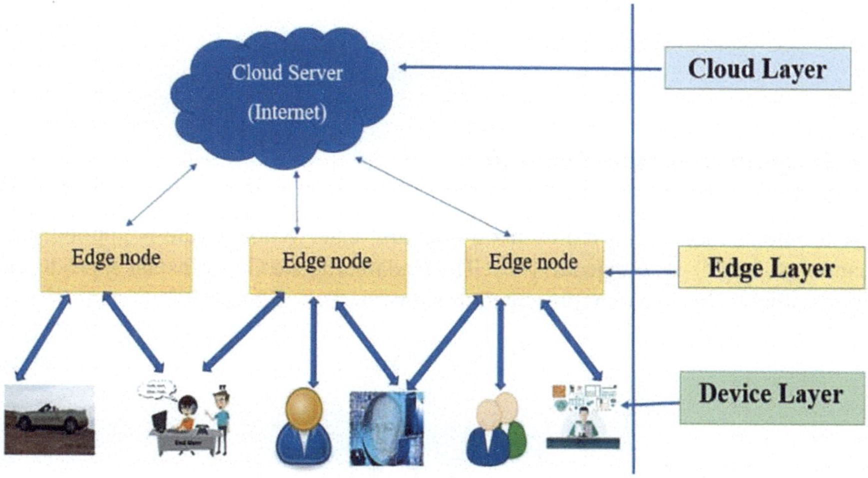

Fig. 7.1 Basic paradigm of edge computing

processes that distinguish it from classical computing. The idea of superposition, in which a qubit can exist in a combination of classical states controlled by probability amplitudes, is captured by the quantum state representation. These states are manipulated to carry out calculations by quantum gate operations, which are represented by unitary matrices. Gates like as the Hadamard create superposition incorporates the way that quantum states collapse into specific outcomes according to their probabilities as is described by measurement equations. The strong correlations between qubits are highlighted by entanglement, which is represented by Bell states. This allows quantum systems to act in ways that are not possible in classical systems. Grover's search strategy increases the probability of effectively locating answers in an unsorted database by utilizing amplitude amplification. When combined, these equations show how quantum systems work and pave the way for advances in computation and problem-solving. These formulas explain how qubits are moved, measured, and employed in quantum algorithms to accomplish computational tasks that are the cornerstone of quantum computing.

7.1.1 Quantum State Representation and Quantum Superposition

Quantum superposition is a core concept in quantum mechanics, where a quantum system can be in multiple states at the same time until it is measured. In case of a quantum bit, it can be a superposition of the classical states $|0\rangle$ and $|1\rangle$. The probability amplitudes are represented by the coefficients $\acute{a}$ and $\acute{e}$, whose squared magnitudes indicate the likelihood of measuring the qubit in either state. This can be represented as shown in Eq. 7.1.

$$|¥\rangle = \acute{a}|0\rangle + \acute{e}|1\rangle, \text{ where } |\acute{a}|^2 + |\acute{e}|^2 = 1 \tag{7.1}$$

7.1.2 Quantum Gate Operation

V is a unitary matrix that represents a quantum gate. When a gate is applied, the quantum state $|¥\rangle$ is changed to $|¥'\rangle$. The Hadamard gate G, for instance, produces superposition. This can be represented as shown in Eqs. 7.2 and 7.3.

$$|¥'\rangle = V|¥\rangle \tag{7.2}$$

$$G = \frac{1}{\sqrt{2}}\begin{pmatrix} 1 & 1 \\ 1 & -1 \end{pmatrix} \tag{7.3}$$

7.1.3 Measurement

The formula for the probability of measuring a quantum state $|¥\rangle$ in a basis state $|j\rangle$ as illustrated in Eq. 7.4.

$$Q(j) = |\langle j|¥\rangle|^2 \tag{7.4}$$

This probability determines which basis state the quantum state will collapse into during measurement.

7.1.4 Entanglement

For a system with two qubits, the following Eq. 7.5 becomes relevant:

$$|¥\rangle = \frac{1}{\sqrt{2}}(|00\rangle + |11\rangle) \tag{7.5}$$

This is an illustration of a Bell state, which denotes maximal entanglement and occurs when one qubit's state is reliant on another.

7.1.5 Grover's Search Algorithm

A reflection operator is used in the amplitude amplification step as illustrated in the following equation:

$$V_s = 2|¥\rangle\langle ¥| - J \tag{7.6}$$

In an unsorted database, this process increases the likelihood of the desired result. Figure 7.1 shows the fundamental architecture of edge computing where Fig. 7.2 indicates the Three-layer architecture of Hybrid Quantum-Edge Computing.

The organization of this chapter is as follows: the basic ideas of quantum edge computing and its motivation are presented in Section 7.2 along with its importance in contemporary computing paradigms. Subsequently Section 7.3 discusses the applications of *QEC* which is followed by the discussion on its challenges in Section 7.4. We have discussed the concept of quantum cloud computing in Section 7.5 while collaborating the security ultimatums in Section 7.6. Afterwards, the privacy technique preservation is detailed in Section 7.7. The next section contains the trust establishment mechanisms in *QEC*. Finally, section 7.9 concludes the paper while suggesting possible directions for future research.

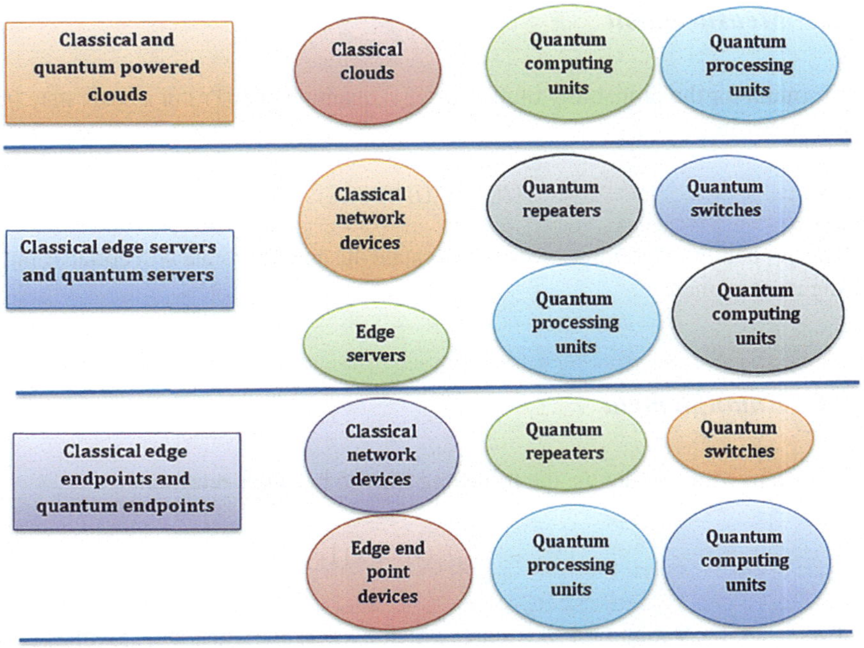

Fig. 7.2 Three-layer architecture of hybrid quantum-edge computing

7.2 Motivation

Decentralized computing paradigms like edge computing have been made possible by the quick development of technology, especially due to the rampant growth of Internet of Things (IoT) devices and the explosion of data produced at the network's edge. By moving computational resources closer to data sources, edge-driven intelligence (EDI) has emerged as a key component of contemporary technology, thereby allowing for real-time data processing, decreased latency, and enhanced scalability. This paradigm change tackles important problems in latency-sensitive applications like industrial IoT systems, healthcare diagnostics, and driverless cars. However, these benefits also come with serious security, privacy, and trust issues. Because edge settings are dynamic and diverse, and because sensitive data processing is becoming more and more important, the ecosystem is vulnerable to threats including denial-of-service assaults, illegal access, and data breaches. For EDI technology to be adopted securely and smoothly, these issues must be resolved. The incorporation of quantum computing into edge frameworks represents a revolutionary advancement in current era of technology. There is revolutionary potential for edge computing applications due to quantum computing's unmatched capacity to tackle intricate computational problems tenfold quicker than classical systems. Quantum computing has the potential to transform predictive analytics, artificial intelligence, and data security by utilizing concepts like superposition, entanglement, and quantum interference.

For example, Quantum key distribution (QKD) solves important security issues in edge situations by offering unbreakable encryption [11]. Additionally, the emergence of edge QEC opens up opportunities for high-precision, real-time calculations in a variety of industries, including finance, defence, healthcare, and smart cities. Although QEC has the potential to revolutionize the field, it also carries with it and intensifies many of the operational and security issues that are present in both edge and quantum computing paradigms.

The pressing necessity to close the security holes in edge quantum computing is what spurred this research. Although quantum computing has strong security features, when combined with edge networks and classical systems, it produces hybrid environments that are susceptible to various threats. These include interoperability problems, classical system exploits, and quantum-specific assaults. Maintaining data integrity and privacy across hybrid infrastructures requires secure communication [12]. It is impossible to overestimate the significance of strong trust management systems, safe device authentication, and cutting-edge cryptographic techniques. Furthermore, creative solutions for scalability, energy economy, and reliability are required due to the early-stage nature of quantum hardware and the technical limitations of deploying quantum systems at the edge. The purpose of this research is to address these issues and offer solutions for building a secure and robust edge quantum environment. This entails investigating methods for protecting privacy, safe hybrid architectures, and trust-building strategies that can support the upcoming edge-driven intelligence generation. By examining these aspects, the study hopes to stimulate QEC developments and promote an innovative, secure, and dependable atmosphere in a digital world that is becoming more linked and decentralized [13].

7.3 Applications of Quantum Edge Computing

By offering quicker, more effective, and extremely secure computational capabilities, edge quantum computing has the potential to completely transform a wide range of sectors. One of the most well-known uses is in artificial intelligence and machine learning, is where deep learning model training and inference can be significantly accelerated by quantum-enhanced algorithms. Large-scale datasets, for example, can be optimized by quantum computing, facilitating the quicker development of real-time decision-making systems, predictive analytics, and more precise natural language processing tools. This capacity is particularly helpful in fields where quick and accurate decision-making is essential, such as autonomous systems. Secure communications is a key additional use in this context. A fundamental component of quantum communication and QKD uses the laws of quantum mechanics to guarantee the secure transmission of encryption keys and guards against illegal interception. From financial transactions to military communications, enterprises may protect sensitive data sent over networks by implementing QKD in edge quantum computing platforms. Furthermore, by instantly detecting and reducing risks, edge quantum computing might improve cybersecurity protocols. Another area where edge quantum computing has the potential to be advantageous is healthcare. Real-time analysis of enormous volumes of medical data can be facilitated by integrating quantum algorithms with edge devices. This can lead to speedier diagnostic procedures, more individualized treatment plans, and advances in drug discovery [14]. Edge servers with quantum capabilities can process genetic data more quickly, while advancing genomics and precision medicine. Furthermore, healthcare systems' logistics can be improved by edge quantum computing, which will enhance patient care and resource allocation.

The way autonomous cars process and respond to sensor data can be revolutionized by edge quantum computing. Vehicle safety and economy can be improved by using quantum-enhanced optimization algorithms to navigate complicated surroundings more accurately. For example, quantum edge computing-powered traffic control systems can handle data from hundreds of vehicles in real time, averting accidents and easing traffic. The broad use of autonomous mobility systems depends on these features. Another exciting use case for edge quantum computing is smart cities. Municipalities may maximize public transit, energy management, and urban planning by utilizing quantum-powered edge devices. Large datasets produced by sensors placed across a city can be analyzed by quantum algorithms, allowing for real-time modifications to infrastructure management and resource allocation. Edge quantum computing, for instance, can improve smart grid efficiency, cutting down on energy waste and consumer expenses [15].

The developments in quantum edge computing have potential benefits for the finance industry as well. Quantum algorithms can be used by financial organizations to improve fraud detection, risk assessment, and portfolio management. These organizations can do real-time analytics on enormous financial datasets and spot trends and abnormalities with previously unheard-of speed and accuracy by integrating quantum processing units at the edge. Furthermore, simulation and scientific research can benefit greatly from edge quantum computing. Simulations in domains like astrophysics, climate modeling, and materials science can be accelerated using quantum-powered edge systems. For example, scientists can forecast how novel materials will behave in certain scenarios or model intricate chemical reactions using edge quantum computing [16]. Innovations in advanced manufacturing, environmental preservation, and renewable energy technology may result from these developments. Edge quantum computing has the potential to improve agricultural productivity and resource management in agriculture. Data from edge sensors placed in fields can be analyzed by quantum algorithms to reveal information on crop health, weather patterns, and soil characteristics. In order to maximize productivity and reduce waste, farmers can use this information to influence data-driven decisions that support sustainable farming methods. Another area where edge quantum computing shows promise is in defense and aerospace applications. Quantum-powered edge systems can enhance situational awareness, enabling faster and more accurate analysis of battlefield data. By integrating quantum algorithms with edge devices, military forces can improve decision-making processes, optimize logistics, and enhance communication security [17]. In the aerospace sector, quantum edge computing can facilitate the design and testing of advanced propulsion systems, materials, and navigation technologies.

Finally, the entertainment and media industries can leverage edge quantum computing to deliver immersive experiences to users. Quantum algorithms can optimize content delivery networks, reducing latency and improving the quality of streaming services. Furthermore, by enabling the real-time representation of complex landscapes, edge quantum computing can improve the realism of applications in virtual and augmented reality [18]. Edge quantum computing is a revolutionary technology with wide-ranging effects due to its adaptability which is positioned as a crucial enabler for the upcoming generation of technological developments due to its capacity to handle computationally demanding operations while retaining low latency and excellent security. Following are the few application areas:

- ***Machine Learning and Artificial Intelligence***: Artificial intelligence (AI) with quantum enhancement can speed up machine learning model training and inference, there by enhancing fields like computer vision, natural language processing, and predictive analytics, etc.
- ***Secure Interactions***: In edge situations, Quantum Key Distribution (QKD) protects sensitive data by ensuring secure encryption key delivery.
- ***Healthcare***: Advanced medication discovery, quicker diagnosis, and individualized therapies are made possible by real-time medical data processing that are made possible by edge quantum computing [19].
- ***Self-Driving Automobiles***: Edge quantum computing improves autonomous vehicle decision-making by processing data quickly and locally, guaranteeing efficiency and safety.
- ***Smart Cities***: Smart city initiatives are supported by quantum-enhanced edge computing, which optimizes traffic control, energy management, and resource allocation.

7.4 Challenges and Limitations of QEC

The innovative concept of quantum edge computing, which blends quantum and edge computing, has enormous potential to transform a variety of sectors, including artificial intelligence and telecommunications. By combining the real-time capabilities of edge computing with the processing power of quantum computers, this hybrid technique allows for safe and effective data processing near the data source. Notwithstanding its potential, quantum edge computing has a number of obstacles and restrictions that must be overcome before it can be widely used. These topics are covered in full in this document. Edge driven intelligence has unique problems that impact the effectiveness, security, and scalability of edge intelligence systems. Edge-driven intelligence focuses on processing and analyzing of data at the network's edge, closer to its origin. Edge driven intelligence has benefits such as reduced latency, improved privacy, and real-time processing capabilities, but it also brings with it drawbacks in terms of resource constraints, data management, security, interoperability, and scalability [20–22].

To fully realize the potential of edge- driven intelligence, overcoming these obstacles requires industry stakeholders to work together, come up with creative solutions, and use durable solutions. With its revolutionary combination of two game-changing technologies, edge quantum computing promises real-time data processing and previously unheard-of computational power. But it also has many obstacles and limitations, ranging from network and hardware limitations to security and financial ones. Policymakers, industry stakeholders, and researchers will need to work together to address these problems. The potential of edge quantum computing can be realized with consistent investment and innovation, opening the door to a new age of technological development.

7.4.1 Data Management and Quality

It might be difficult to integrate and manage data from several heterogeneous edge devices, which increases the risk of data inconsistencies and quality problems. It might be difficult to guarantee prompt and accurate real-time data stream processing, particularly when dealing with different data quantities and velocities.

7.4.2 Latency and Network Dependence

Real-time decision-making is impacted by communication delays between edge devices and central systems, even with edge computing's reduction of latency. Applications that use edge-driven intelligence can't function well if there are unreliable network connections that stop data from flowing.

7.4.3 Energy Efficiency

Efficient processing and communication solutions are essential for edge devices since they frequently function in low power settings. Guaranteeing longer battery life for edge devices is extremely challenging, especially when those devices are in remote or mobile situations.

7.4.4 Real-Time Decision Making

Because of potential synchronization and latency issues, it is challenging to accomplish real-time decision- making in edge environments. Complex coordination methods are needed to guarantee consistent decision- making across several edge nodes.

7.4.5 Limitations in Hardware

The processing capability of edge devices is often lower than that of centralized data centers, which makes complex data analysis and machine learning tasks difficult to accomplish. Limitations in memory and storage on edge devices might make it difficult to handle big datasets and sophisticated models. Integrating quantum computers into edge systems is difficult since they need strict environmental conditions, such intense cooling is required. The foundation of quantum computing, quantum processors are still in their infancy. The number of qubits that can be handled by current quantum computers is limited, and the qubits are very prone to errors from environmental noise. One of the biggest challenges is creating reliable, scalable, and fault-tolerant quantum processors that can be used in edge environments. For a number of tasks, edge computing significantly depends on classical systems. The smooth integration of quantum processors with classical systems necessitates complex hybrid architectures, which are still being researched. It is crucial to provide effective communication between quantum and classical systems without a lot of latency or data loss [23–25]. Cryogenic settings are frequently necessary for quantum computers in order to preserve qubit stability. One major challenge is deploying such systems at the edge, where space and resources are few. It would be revolutionary to create quantum processors that can function at room temperature or in less constrained settings.

7.4.6 Networks Challenges

The success of edge quantum computing systems depends on the smooth integration of classical and quantum communication connections. Strong quantum communication lines are necessary for edge quantum computing in order to transfer quantum states and entanglement between nodes. Photon loss, noise, and short transmission range are some of the limitations that affect quantum communication. Now-a-days, trustworthy nodes or repeaters are frequently needed for quantum networks, which can jeopardize scalability and security. Complicating matters is the combination of quantum and classical communication networks [26]. Despite the established nature of classical networks, new standards and protocols must be developed to integrate them with quantum systems. It is a challenging ultimatum to guarantee efficiency and compatibility in such hybrid networks. Although the goal of edge computing is to reduce latency, the time needed for quantum operations and error correction may result in additional delays when quantum computing is included. Furthermore, current communication infrastructures may be strained by the bandwidth needed for transmitting quantum information.

7.4.7 Scalability and Deployment Issues

An ongoing difficulty is creating scalable solutions for quantum edge networks, which call for reliable error correction and quantum entanglement distribution protocols. Investing heavily in infrastructure, such as power, cooling, and physical space, is necessary to deploy quantum computing resources at the edge. One major technical problem is scaling quantum systems to accommodate large-scale edge installations. Edge situations sometimes require large-scale, distributed computing operations that are currently beyond the capabilities of current quantum technology. Scalability depends on integrating quantum technologies into current edge ecosystems and guaranteeing compatibility across manufacturers. However, this approach is hampered by the lack of uniformity in quantum software and hardware.

7.4.8 Security and Privacy Issues

Despite the inherent security of quantum communication, there are hazards associated with the hybrid architecture that must be addressed due to flaws in traditional systems. Although quantum communication offers unmatched security, edge quantum computing's hybrid design creates risks. Traditional cyber attacks can still target traditional components of the system, there by jeopardizing the system's overall security. Because edge devices are frequently placed in isolated or unprotected areas, they are susceptible to theft or physical alteration. One of the biggest challenges is making sure that quantum-enabled edge devices are physically secure [27]. It is difficult to maintain data privacy in a hybrid setting where quantum and classical systems coexist. Even in the face of sophisticated quantum threats, quantum systems must guarantee the security of sensitive data processed at the edge.

7.5 Quantum Cloud Computing

The emergence of quantum cloud computing has the potential to completely transform the computational environment by providing previously unheard-of possibilities for resolving some of the most challenging and resource-intensive issues in a variety of fields. For many years, classical computing has served as the foundation for technological progress; nonetheless, it faces considerable challenges in solving issues such as large-scale optimization, cryptographic algorithms, molecular simulations, and Artificial Intelligence (AI) model training. The investigation of quantum computing, which uses concepts like superposition, entanglement, and quantum interference to produce exponential computational capabilities has been spurred by these difficulties. Quantum cloud computing democratizes access to these potent computational resources by fusing cloud infrastructure with quantum computing [28]. This enables researchers, developers, and organizations to take advantage of quantum capabilities without having to make large investments in specialized quantum hardware. As data-driven businesses increase and computation-intensive jobs become more widespread, this paradigm is ideally suited to meet the increasing demand for decentralized, scalable, and effective computing platforms. The potential of quantum cloud computing to overcome the drawbacks of traditional computing is one of its main drivers. Traditional algorithms frequently fall short of offering effective solutions in reasonable amounts of time as datasets get bigger and more complicated. These procedures can be greatly accelerated by quantum cloud computing, opening the door to innovations in fields including financial analysis, medicine development, climate modelling, and cryptography. For example, Grover's method speeds up search operations, whereas Shor's algorithm is a quantum algorithm that has the ability to crack traditional cryptographic protocols [29]. This revolutionary ability promises to open up completely new possibilities that were previously thought to be unattainable in addition to solving current computational issues.

Furthermore, accessibility and inclusivity are guaranteed by the cloud-based distribution architecture of quantum computing. Superconducting qubits and trapped-ion systems are examples of quantum gear that is extremely costly to build and maintain and necessitates extremely regulated settings. It is not feasible for the majority of organizations and researchers to invest in such infrastructure. This problem is addressed by quantum cloud platforms, which offer quantum computing as a service (QCaaS) and are offered by firms such as IBM, Google, Microsoft, and Amazon. With the help of this service model, where consumers can access state-of-the-art quantum processors online without having to pay high upfront or ongoing maintenance fees [30]. This makes it possible for businesses, academic institutions, and individual researchers to engage in quantum innovation, which advances the discipline and promotes a worldwide quantum ecosystem.

Quantum cloud computing not only democratizes access but also fosters interdisciplinary cooperation and creativity. Researchers and developers from different disciplines can collaborate on quantum algorithms and applications by offering a common platform. Tools for coding, debugging, and simulating quantum programs are frequently included in cloud-based quantum development environments, which facilitate learning and encourage innovation. Hybrid quantum-classical approaches to problem-solving are also made possible by these systems' smooth integration with classical computer resources. Given that present quantum systems are still in their infancy and need classical counterparts for preprocessing, error correction, and data interpretation, this hybrid model is particularly useful in the near future [31].The potential of quantum cloud computing to revolutionize data security and privacy which stands as another important driving force. Although quantum computers pose a danger to conventional encryption approaches, they also open the door for next-generation cryptographic strategies like post-quantum cryptography and quantum key distribution (QKD).

Cloud platforms with quantum capabilities can be used as test sites for putting these new security policies into practice and assessing their effectiveness. Quantum cloud computing provides a proactive method of protecting sensitive data as cyber attacks get more complex, thereby helping businesses maintain their lead in the cybersecurity arms race. Quantum computing's incorporation into cloud systems is also consistent with the more general trends of energy efficiency and sustainability. Particularly for certain problem sets, quantum systems may be able to execute computations with a substantially lower energy usage than traditional supercomputers. Cloud data centres can improve energy consumption and lower the carbon footprint associated with distributed hardware deployments by centralizing quantum resources. With governments and businesses throughout the world pledging to meet aggressive climate targets, this sustainability feature is especially becoming alluring. Quantum cloud computing is also essential for education and workforce development. Quantum computing is a very specialized field that calls for knowledge of mathematics, computer science, and physics. Professionals and students can experiment with actual quantum devices and simulations. These platforms foster the next generation of quantum scientists and engineers by providing extensive documentation, tutorials, and community forums. Quantum cloud computing guarantees a consistent flow of personnel prepared to advance the area by reducing entry barrier [32–34].

Quantum cloud computing has enormous potential, but it also has drawbacks that spur more study and development. The full potential of quantum systems is nevertheless hampered by problems like qubit coherence, error correction, and algorithm optimization. Continuous changes of quantum hardware and software are made possible by the cloud model, which offers a workable framework for resolving these issues. It also makes it easier for a worldwide user base to provide real-time feedback, which speeds up the development of quantum technologies and promotes an iterative improvement culture. Applications of quantum cloud computing are found in many different industries, each with its own incentives and opportunities. By simulating molecular interactions at a scale never before possible, quantum algorithms have the potential to transform drug discovery in the healthcare industry and accelerate the development of life-saving therapies. Quantum computing has unmatched accuracy in detecting fraud, managing risks, and optimizing portfolios in the financial industry [35].Quantum solutions in logistics help to optimize supply chains, lowering expenses and boosting productivity. These applications highlight how revolutionary quantum cloud computing is establishing it as a key component of upcoming developments. The need to transcend traditional computational constraints democratize access to cutting-edge technology, improve cooperation, and tackle urgent global issues are some of the many factors that motivate quantum cloud computing. Quantum cloud computing is a paradigm leap in computation by fusing the scale and

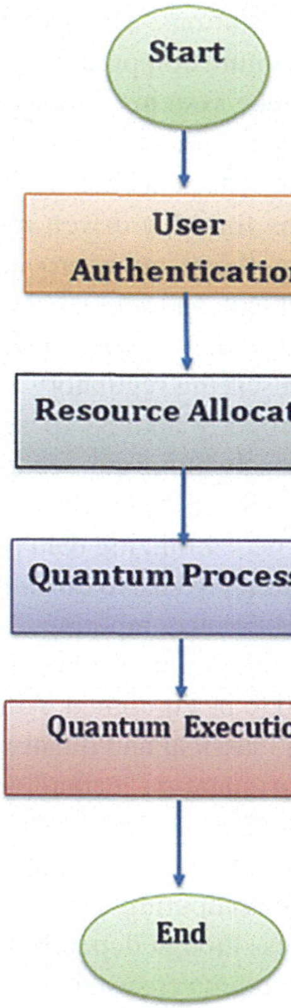

Fig. 7.3 Work flow of quantum cloud computing

accessibility of cloud platforms with the power of quantum physics. Its significance as a revolutionary force is highlighted by its potential to propel scientific, industrial, and educational advances, encouraging further investment, research, and innovation in the field [36]. The incorporation of quantum computing into cloud ecosystems will surely be crucial in determining how humanity develops technologically as quantum technologies advance. Figure 7.3 shows a scenario of work flow of quantum cloud computing.

7.6 Security Challenges in QEC

Quantum edge computing is the combination of edge and quantum computing, is a revolutionary development in the field of technology. This hybrid paradigm overcomes the latency and bandwidth limitations of centralized cloud computing while providing unmatched computational power and faster processing rates. To protect data and computing operations, it also presents difficult security issues that call for creative solutions. This paper examines the different security issues that come with edge quantum computing and suggests solutions in various domains. In edge-driven intelligence systems, threats to security might originate from a number of sources, including hostile people, compromised technology, and unprotected communication channels. The scattered and heterogeneous nature of edge settings presents special problems for edge-driven intelligence security, necessitating a complete solution. Strong authentication procedures, secure communication protocols, timely patch management, and sophisticated intrusion detection systems are a few examples of this [37]. Trust and security in edge-driven intelligence applications are also contingent upon preserving regulatory compliance and safeguarding data privacy.

To properly address these challenges, industry players must work together and continuously innovate security practices [38]. Edge computing places computer resources nearer to the data sources, like local servers and Internet of Things devices, thereby decentralizing data processing. However, quantum computing uses the laws of quantum physics to do complicated computations at exponentially quicker speeds than traditional computing. By combining these two paradigms, the processing capability of quantum systems and the low-latency advantages of edge computing are intended to be combined. Although this hybrid approach shows promise, the coexistence of classical and quantum systems, edge network vulnerabilities, and the early stages of quantum technology provide special security challenges. These difficulties include safe communication between hybrid infrastructures, privacy issues, and data integrity. By combining the advantages of edge and quantum computing, edge quantum computing has the potential to completely transform a number of sectors. Its success though depends on resolving important security issues, building a robust edge quantum ecosystem requires addressing quantum communication weaknesses, protecting data privacy, securing devices, and integrating hybrid systems. Investing in hardware security, trust frameworks, research, and sophisticated cryptography will be essential to risk mitigation. The promise of edge quantum computing can be achieved without jeopardizing data integrity and privacy by taking a proactive approach to security, opening the door to a safe and creative future [39]. The following are some of the main security issues with edge- driven intelligence:

I. **Quantum Communication Vulnerabilities**: Data transmission in quantum communication depends on quantum states like qubits. Based on the ideas of quantum physics, QKD provides potentially indecipherable encryption; yet, assaults can be launched against real-world implementations. Among the examples are: Attacks such as "photon number splitting" use multi-photon emissions to intercept communications. Intercepting traditional communication lines necessary for QKD protocol negotiations is known as a "man-in-the-middle" attack.

II. **Privacy of Data in Edge Settings**: By processing and storing data close to its source, edge computing raises the possibility of data breaches and illegal access. These dangers are made worse in hybrid systems by the difficulty of combining quantum and classical data streams. Among the difficulties are: Combining private information from several edge devices may result in privacy infringement. Quantum Data protecting the privacy of quantum data while it's being processed and sent.

III. **Data and Device Security**: Sensitive information is frequently stored locally on edge devices where it may be unintentionally exposed if not improperly encrypted. Strong encryption techniques are crucial because unprotected data transferred between edge devices and central systems or other devices can be intercepted.

IV. **Network Security**: Attacks such as man-in-the-middle, eavesdropping, and data manipulation can be carried out through insecure communication protocols. Adversaries may be able to travel laterally from compromised edge devices to different areas of the network if the network is not properly divided.

V. **Authentication and Authorization**: Sensitive information and systems may be accessed by unauthorized parties through edge devices with weak or default authentication methods. An environment that is distributed makes it difficult to establish fine-grained access control regulations, which raises the possibility of illegal access and possible data breaches.

VI. **Integrity and Availability**: Erroneous judgments and actions can be caused by compromised data; hence it is critical to maintain the integrity of data processed and stored at the edge. Denial-of-service (DoS) attacks can cause major disruptions to edge devices, particularly in real-time applications, by impairing their functioning and availability.

VII. **Intrusion Detection and Prevention**: It is difficult to identify anomalies and possible intrusions in a dispersed edge environment; this calls for sophisticated analytics and monitoring tools. It is challenging to plananefficient response to security issues that include multiple edge devices and locations.

VIII. **Data Privacy**: Edge devices handle sensitive and private data often, which calls for strict privacy controls to abide by laws and safeguard user information. The risk of privacy breaches can be decreased by putting data minimization principles into practice and collecting and processing just the information that is required.

IX. **Supply Chain Security**: It is essential to make sure that the software and hardware utilized in edge devices are from reliable sources and haven't been tampered with throughout the supply chain. Using third-party software and components adds further security concerns that must be properly handled.

X. **Trust Management**: The security and integrity of data processed and transferred at the edge must be maintained to prevent manipulation and unauthorized access. Secure boot methods and firmware update accuracy must be verified in order to shield edge devices from potential compromises.

7.7 Privacy Preserving Techniques in Quantum Edge Driven Intelligence

With the sensitive nature of the data gathered from Internet of Things devices, privacy becomes a critical consideration in edge-driven intelligence systems. Data protection rules must be observed, and privacy concerns can be reduced by using privacy-preserving strategies. Edge-driven intelligence offers advantages including lower latency and increased efficiency by processing data closer to its source. Significant privacy problems are brought up by this decentralization, nevertheless. To safeguard sensitive data and taking use of edge computing's benefits, privacy-preserving measures must be put into practice. Protecting sensitive data in edge-driven intelligence requires the use of privacy-preserving techniques [40]. Further developments in privacy-preserving technologies will be necessary as edge computing develops in order to handle new issues and guarantee strong data protection. Several commonly employed privacy-preserving methods consist of the followings:

(i) **Data Anonymization and Pseudonymization**
In order to prevent individual identification, this method entails eliminating Personally Identifying Information (PII) from datasets. Data suppression, generalization, and anonymization are common techniques. The analysis utility of the dataset is maintained while individual identities are protected by substituting false identifiers, or pseudonyms, for private identifiers.

(ii) **Federated Learning**
Federated learning does not require transferring data to a central server; instead, it leverages machine learning models use local data from several edge devices. Privacy is protected by ensuring that raw data stays on the relevant devices. The only information sent to a central server is model updates, like gradients, which are combined to update the global model and improve data privacy even more.

(iii) **Differential Privacy**
In order to mask the impact of each one data point, differentiating privacy entails introducing random noise to the data or the analysis findings which are emerging to be crucial. Private information cannot be easily deduced from the output because of this. With the quantitative privacy guarantees provided

by differential privacy, businesses may effectively weigh the benefits and drawbacks of data privacy and utility.

(iv) **Homomorphic Encryption**
Because calculations on encrypted data can be completed without first requiring decryption, homomorphic encryption protects data confidentiality during processing. Advances in homomorphic encryption, despite its computational intensity, are making it more practical for application in edge devices, allowing for data analysis that protects privacy.

(v) **Trusted Execution Environments (TEEs)**
TEEs offer a safe haven inside a processor where private information can be handled separately and shielded from other system components. TEEs provide the security and integrity of data throughout processing, protecting it against manipulation and unauthorized access.

(vi) **Local Differential Privacy (LDP)**
To ensure that data is anonymized before it is transferred or processed, LDP utilizes differential privacy techniques locally at the data source (for example, on the edge device).Even if aggregated data is studied, LDP techniques aid in preventing the reconstruction or identification of specific user data.

(vii) **Edge-Based Privacy Policies**
Enforcing and implementing privacy rules at the edge guarantees that organizational policies and privacy regulations are followed when handling data. Strict enforcement of access control mechanisms is essential to limit sensitive data access to only authorized entities adhering to predetermined policies.

(viii) **Block Chain Technology**
By eliminating the need for a central authority and offering a decentralized framework for transparent and safe data management, block chain can guarantee data integrity and traceability. Innovative block chain techniques such as zero-knowledge proofs, can preserve anonymity while enabling data sharing and transactions in edge scenarios.

7.8 Trust Establishment Mechanisms in QEC

Creating a foundation of trust for quantum edge-driven intelligence requires a mix of techniques and tools for identity verification, data integrity, and secure communication. In distributed and decentralized edge contexts, trust- building mechanisms like zero trust architecture, reputation-based systems, block chain, PKI and TEE s are essential. The security and dependability of edge-driven intelligence systems depend on the efficient implementation of these mechanisms [41]. Establishing trust between entities is crucial for promoting cooperation and team work in edge-driven intelligence systems. Ensuring the liability and authenticity of edge nodes and services requires mechanisms for building trust. Establishing confidence in edge-driven intelligence is necessary in decentralized settings to maintain the security, dependability, and integrity of data and services. Trust between edge devices, users, techniques for establishing trust include the following:

(i) **Device Authentication**
By putting strong authentication techniques in place, such as biometrics, multi- factor authentication (MFA), or public key infrastructure (PKI), itis ensured that only authorized devices can connect to the network. A device's identity may be verified and ensure it isn't malicious or compromised by utilizing digital certificates from a reliable certificate authority (CA).

(ii) **Secure Boot and Firmware Integrity**
Secure boot procedures make guarantee that only reliable software loads on a device by confirming the boot loader's and operating system's digital signatures. Firmware integrity is regularly verified using cryptographic signatures or checksums to make sure it hasn't been tampered with.

(iii) **Trusted Execution Environments (TEEs)**
TEEs offer a safe haven where private information which can be handled discreetly and separately from other system functions inside the device's CPU. By ensuring the security and integrity of data while it is being handled, TEEs protect against illegal access and modification.

(iv) **Blockchain and Distributed Ledger Technologies**
Blockchain technology provides an unchangeable and transparent transaction ledger, which can provide a decentralized framework for edge device trust-building. Agreements and trust policies between parties can be automatically enforced and automated on a block chain without requiring a central authority by using smart contracts.

(v) **Zero Trust Architecture**
The foundation of zero trust architecture is the notion that no entity, whether internal or external to a network, should ever be trusted by default. Maintaining confidence and lowering the risk of breaches can be achieved by routinely confirming and validating the identification and security posture of devices and users.

(vi) **Identity and Access Management (IAM)**
Ensuring that only authorized and authenticated entities are able to access resources is a key function of robust Identity and Access Management (IAM) systems. Sensitive data is protected from unauthorized access by establishing and implementing access control policies based on roles, attributes, and contextual data.

(vii) **Public Key Infrastructure (PKI)**
PKI ensures secure communications by using digital certificates for device and system authentication and trust-building. Trust must be upheld by using efficient key management procedures, which include the creation, sharing, and revocation of cryptographic keys.

(viii) **Data Integrity and Provenance**
Data integrity is ensured by using cryptographic hashing, which makes sure that data is not changed while being transmitted or stored. Data integrity

and authenticity can be established by following its origin and history, which guarantees that the information originates from reliable sources.

(ix) **Collaborative Trust Models**
Mutual authentication builds confidence since it requires both parties to a communication to verify each other's identities. By creating trust federations, disparate organizations can acknowledge and have faith in one other's users and devices because of common agreements and policies and central systems can be established and maintained through a variety of strategies. The important academic contributions that explore the complex aspects of security, privacy, and trust in the context of quantum edge-driven intelligence are comprehensively and perceptively summarized in the given Table 7.1. Each post skillfully reveals the goals, creative solutions, and problems that eminent writers have tackled. It gives readers a broad overview of the most recent innovations and tactical approaches by weaving a rich tapestry of information that illustrates the progress and breakthroughs that define this cutting-edge field. Table 7.2. shows the comparative analysis between edge computing and Cloud computing.

7.9 Conclusion and Future Scope

This chapter provides a thorough overview of the security, privacy and trust challenges pertaining to quantum edge- driven intelligence systems. It is possible to reduce risks and create resilient edge settings that encourage confidence and trust among users and stakeholders by utilizing the right strategies and procedures. Trust, security, and privacy are critical factors that must be taken into account when creating and implementing quantum edge-driven intelligence solutions. Ensuring the confidentiality, integrity and availability of data becomes more difficult yet crucial as data processing gets closer to the point of generation. The security, privacy and trust issues surrounding quantum edge-driven intelligence systems are covered in detail in this chapter. By using the appropriate techniques and procedures, it is possible to lower risks and build resilient edge settings that promote confidence and trust among users and stakeholders. When developing and putting into practice edge-driven intelligence solutions, trust, security, and privacy are vital considerations. As data processing moves closer to the point of generation, maintaining data availability, confidentiality, and integrity becomes more challenging but also more important.

Table 7.1 An overview of current studies on trust, security, and privacy in edge-driven intelligence

Authors/Year	Research focus	Objective/Purpose	Key contributions	Challenges addressed	Solutions
Shi et al. [5]	Edge Computing	To explain edge computing and some of its possible advantages	Defined edge computing, emphasizing latency reduction by processing data closer to the source	Latency and bandwidth limitations in cloud-centric models	Offloading data processing tasks to edge devices for faster response times
Zhang et al. [13]	Protecting Privacy through Edge Data Aggregation	To create techniques for dispersed edge settings' safe data aggregation	Suggested safe aggregation strategies that use cryptographic techniques to safeguard privacy in edge federated learning	Hazards to privacy in edge environments when data is aggregated	Methods for safe multi-party computing and homomorphic encryption
Satyanarayanan [24]	Security and Trust in Mobile Edge Technology	To explore trust establishment between edge nodes and mobile devices	Emphasized safe offloading techniques and the significance of fostering trust between edge nodes and devices	Offloading tasks securely and managing trust	Frameworks for authentication and protocols for safe offloading
Li et al. [43]	Secure IoT Data Transmission in Edge Computing	To ensure confidentiality and integrity of data shared among IoT devices in edge setups	Lightweight cryptographic algorithms were used to provide encryption-based frameworks	Data exchange without jeopardizing data privacy	Simple encryption techniques that work well with IoT devices with limited resources
Xu et al. [44]	Federated Learning with Differential Privacy	To integrate differential privacy into federated learning at the edge for secure training	Presented privacy-preserving techniques that guarantee the anonymity of each person's data inputs throughout training	Hazards of privacy leaks when training collaborative models	In federated models, different privacy measures are used to mask individual data contributions
Dinh et al. [45]	Edge Computing Security and Resource Management	To improve secure resource allocation and task scheduling at the edge	Developed techniques for edge networks' safe and effective distribution of processing power	When allocating, the edge resources' availability and integrity	Dynamically scheduled frameworks for safe resource management

(continued)

Table 7.1 (continued)

Authors/Year	Research focus	Objective/Purpose	Key contributions	Challenges addressed	Solutions
Liu et al. [46]	Blockchain to Promote Trust in Edge Computing	To use blockchain for enhancing trust and data integrity in edge environments	Blockchain technology was used to guarantee openness and reliability in communications between edge nodes	In decentralized edge networks, trust and data integrity are crucial	Ledgers built on the blockchain to offer unchangeable records of node-to-node interactions
Zhou et al. [47]	Edge's Safe and Effective AI Model Training	To mitigate security risks in training AI models in edge computing	Researched ways to protect AI training, lowering the danger of hostile attacks on edge AI systems	Problems in security when training distributed AI	Adversarial training and encryption-based data protection
Wang et al. [48]	Maintaining Privacy and Anonymity in IoT Computation	To protect user identities and ensure privacy in IoT communication at the edge	Developed methods for IoT device computation and anonymous communication that protect privacy	IoT networks' secure communication and identity privacy	Frameworks for computation that protect privacy and anonymity protocols
Ren et al. [49]	Edge Computing Security Powered by AI	To leverage AI for detecting and responding to security threats in edge environments	Investigated threat detection systems powered by AI that are able to recognize and react to threats instantly	Threat detection and reaction in real time at the edge	Security threats are dynamically identified and mitigated by machine learning models

Table 7.2 Comparison table of edge computing versus quantum computing

Feature	Edge computing	Quantum computing
Primary focus	Processing data close to the source	Leveraging quantum phenomena for computation
Key components	IoT devices, Edge servers, Network routers	Qubits, quantum processors, Quantum gates
Latency	Low latency for real-time processing	May introduce latency due to quantum operations
Power requirements	Moderate to high, Depending on workload	Extremely high, Often requiring cryogenic environments
Error management	Handled by traditional error-checking mechanisms	Requires complex quantum error correction
Deployment environment	Flexible, Near data sources	Limited to specialized setups (e.g., cryogenic labs)
Security	Relies on encryption and network security	Intrinsically secure via quantum mechanics
Scalability	Well-established and scalable	Limited by current technology and infrastructure
Cost	Relatively lower	Very high due to specialized hardware
Applications	IoT, AI, Real-time analytics	Cryptography, Optimization, Quantum simulation

Acknowledgements The authors are really very appreciative to the Siksha 'O' Anusandhan University for affording the support to make this investigation successful.

References

1. Mian, A.S., et al.: A survey on security, privacy, and trust in edge computing. IEEE Commun. Surv. Tutorials **22**(4), 2673–2725, Fourthquarter (2020)
2. Roman, R., et al.: On the features and challenges of security and privacy in distributed internet of things. Comput. Netw.. Netw. **57**(10), 2266–2279 (2013)
3. Li, H., et al.: Edge computing for internet of things: a survey. IEEE Access **6**, 6900–6919 (2018)
4. Li, G., et al.: A survey on the edge computing for the internet of things. IEEE Access **8**, 6538–6553 (2020)
5. Shietal, W.: Edge computing: vision and challenges. IEEE Internet Things J. **3**(5), 637–646 (2016)
6. Mahindru, R., et al.: Security and privacy in edge computing: a review. In: Proceedings of IEEE 4th International Conference on Computing, Communication and Security (ICCCS), pp. 1–6 (2019)
7. Zhang, H., et al.: Edge computing security: state of the art and challenges. IEEE Internet Things J. **7**(10), 8954–8969 (2020)
8. Atanassov, R., et al.: Privacy-preserving technologies in the era of edge intelligence: a survey. IEEE Trans. Netw. Serv. Manag.Netw. Serv. Manag. **18**(2), 951–966 (2021)

9. Smith, J., Johnson, A.: Enhancing security in edge-driven intelligence: challenges and solutions. IEEE Trans. Edge Comput. **8**(3), 215–228 (2020)
10. Wang, L., Chen, H.: Privacy-preserving techniques for edge-driven intelligence: a survey. IEEE Internet Things J. **6**(5), 7895–7910 (2019)
11. Li, X., Zhang, Y.: Establishing trust in edge-driven intelligence environments: a review. IEEE Trans. Dependable Secure Comput. **18**(2), 143–156 (2021)
12. Gupta, S., Sharma, R.: Future directions and research challenges in security, privacy, and trust in edge-driven intelligence. IEEE Access **10**, 20292–20305 (2022)
13. Zhang, Q., Li, Y.: Security, privacy, and trust in edge computing: a comprehensive survey. IEEE Commun. Surv. Tutorials **20**(4), 3759–3780 (2018)
14. Patel, K., Patel, D.: Edge-Driven Intelligence: A Case Study of Security and Privacy Challenges. IEEE Secur. Privacy Mag. **18**(3), 45–52 (2020)
15. Huang, W., Liu, C.: Trust establishment mechanisms in edge-driven intelligence: current trends and future directions. IEEE Trans. Inf. Forensics Secur. **14**(7), 1796–1809 (2019)
16. Zhao, X., Zhang, W.: Privacy-preserving techniques in edge-driven intelligence: challenges and opportunities. IEEE J. Sel. Areas Commun. **39**(2), 450–463 (2021)
17. Singh, R., Jain, A.: Edge-driven intelligence: challenges and opportunities for security, privacy, and trust. IEEE Potentials **38**(3), 20–25 (2019)
18. Chen, Z., Wang, S.: A survey on security, privacy, and trust in edge-driven intelligence. IEEE Trans. Emerg. Top. Comput. **8**(2), 234–247 (2020)
19. Hagan, M., Siddiqui, F., Sezer, S.: Policy-based security modelling and enforcement approach for emerging embedded architectures. In: 31st IEEE International System-on-Chip Conference (SOCC), pp. 84–89 (2018)
20. Sharma, V., et al.: Security, privacy and trust for smart Mobile-Internet of Things (M-IoT): a survey. CoRR (2019). Available: http://arxiv.org/abs/1903.05362
21. Huang, B., Li, Z., Tang, P., Wang, S., Zhao, J., Hu, H., Li, W., Chang, V.: Security modeling and efficient computation offloading for service workflow in mobile edge computing. Future Gener. Comput. Syst. **97**, 755–774 (2019)
22. Yoon, J.: Trust worthiness of dynamic moving sensors for secure mobile edge computing. Computers **7**(4), 63 (2018)
23. Yuan, J., Li, X.: A multi-source feedback-based trust calculation mechanism for edge computing. In: Proceedings of IEEE INFOCOM-IEEE Conference Computing Communication Workshops (INFOCOM WKSHPS), pp. 819–824 (2018)
24. Satyanarayanan, M.: The emergence of edge computing. Computer **50**(1), 30–39 (2017)
25. Zhouetal, Y.: Edge computing: vision and challenges. IEEE Internet Things J. **4**(5), 637–646 (2017)
26. Bragadeesh, S., Arumugam, U.: A conceptual frame work for security and privacy in edge computing. In: Edge Computing, pp. 173–186. Springer (2019)
27. Esiner, E., Datta, A.: Layered security for storage at the edge: on decentralized multi-factor access control. In: Proceedings of the 17th International Conference on Distributed Computing and Networking, p. 9. ACM (2016)
28. Kang, J., Yu, R., Huang, X., Wu, M., Maharjan, S., Xie, S., Zhang, Y.: Blockchain for secure and efficient data sharing in vehicular edge computing and networks. IEEE Internet Things J. **6**(3), 4660–4670 (2019)
29. Yi, S., Qin, Z., Li, Q.: Security and privacy issues of fog computing: a survey. In: International Conference on Wireless Algorithms, Systems, and Applications, pp. 685–695. Springer (2015)
30. Abawajy, J., Huda, S., Sharmeen, S., Hassan, M.M., Almogren, A.: Identifying cyber threats to mobile-IoT applications in edge computing paradigm. Future Gener. Comput. Syst. **89**, 525–538 (2018)
31. Kozik, R., Chora, M., Ficco, M., Palmieri, F.: Ascalable distributed machine learning approach for attack detection in edge computing environments. J. Parallel Distrib. Comput. **119**, 18–26 (2018)
32. Pyrkov, A., et al.: Quantum computing for near-term applications in generative chemistry and drug discovery. DrugDiscov. Today **28**, 103675 (2023)

33. Mudedla, S.K., Braka, A., Wu, S.: Quantum-based machine learning and AI models to generate force field parameters for drug-like small molecules. Front. Mol. Biosci.Biosci. **9**, 1002535 (2022)
34. Wang, Z., et al.: Improving machine learning force fields for molecular dynamics simulations with fine-grained force metrics. J. Chem. Phys. **159** (2023)
35. Zhang, L., et al.: Editorial: combined artificial intelligence and molecular dynamics (AI-MD) methods. Front. Mol. Biosci.Biosci. **9**, 1012785 (2022)
36. Hasanpour, M., Shariat, S., Barnaghi, P., Hoseinitabatabaei, S.A., Vahid, S., Tafazolli, R.: Quantum load balancing in ad hoc networks. Quantum Inf. Process. **16**, 148 (2017)
37. Nurminen, J.K., Meijer, A., Salmenpera, I., Becker, L.: The next bottleneck after quantum hardware will be quantum software. Ercim. News **128**, 9–10 (2022)
38. Litinski, D.: A game of surface codes: large-scale quantum computing with lattice surgery. Quantum **3**, 128 (2019)
39. Sludds, A., Bandyopadhyay, S., Chen, Z., Zhong, Z., Cochrane, J., Bernstein, L., Bunandar, D., Dixon, P.B., Hamilton, S.A., Streshinsky, M., et al.: Delocalized photonic deep learning on the internet's edge. Science **378**, 270–276 (2022)
40. Thew, R., Jennewein, T., Sasaki, M.: Focus on quantum science and technology initiatives around the world. Quantum. Sci. Technol. **5**, 010201 (2020)
41. Nath, M.P., Priyadarshini, S.B., Mishra, D., Mishra, B.K.: A systematic review on blockchain security technology and big data employed in cloud environment. In: A Roadmap for Enabling Industry 4.0 by Artificial Intelligence, pp. 79–109 (2022)
42. Zhang, K., Ni, J., Yang, K., Liang, X., Ren, J., Shen, X.: Security and privacy in smart city applications: challenges and solutions. IEEE Commun. Mag. **56**(4), 122–129 (2018). https://doi.org/10.1109/MCOM.2018.1700296
43. Li, X., Li, J., Huang, X., Xiang, Y., Zhou, W.: Secure attribute-based data sharing for resource-limited users in cloud computing. Comput. Secur. **72**, 1–12 (2019). https://doi.org/10.1016/j.cose.2017.08.005
44. Xu, Y., Li, Y., Zhang, Y., Sun, L., Qi, H.: A survey on federated learning. IEEE Trans. Neural Netw. Learn. Syst. (2020). https://doi.org/10.1109/TNNLS.2020.3006370
45. Dinh, T.Q., Tang, J., La, Q.D., Quek, T.Q.S.: Offloading in mobile edge computing: task allocation and computational frequency scaling. IEEE Trans. Commun. **65**(8), 3571–3584 (2021). https://doi.org/10.1109/TCOMM.2017.2694425
46. Liu, Y., Yu, F.R., Yin, H., Wang, Y.: Blockchain and machine learning for communications and networking systems. IEEE Commun. Surv. Tutor. **22**(2), 1392–1431 (2020). https://doi.org/10.1109/COMST.2020.2963985
47. Zhou, Z., Chen, X., Li, E., Zeng, L., Luo, K., Zhang, J.: Edge intelligence: paving the last mile of artificial intelligence with edge computing. Proceedings of the IEEE **107**(8), 1738–1762 (2019). https://doi.org/10.1109/JPROC.2019.2918951
48. Wang, C., Zhang, Q., Zhang, H., Liu, J.: Toward privacy-preserving edge computing: challenges and solutions. Comput. Netw. **183**, 107593 (2020). https://doi.org/10.1016/j.comnet.2020.107593
49. Ren, J., Yu, G., He, Y., Wang, Y.: Collaborative edge computing for AI-enhanced smart city services: a review. IEEE Internet of Things J. **9**(1), 379–397 (2022). https://doi.org/10.1109/JIOT.2021.3075191

Chapter 8
Quantum-Inspired Aspect-Based Sentiment Analysis Using Natural Language Processing

Mrinmoy Kayal, Mohinikanta Sahoo, Jayadeep Pati, and Ranjan Kumar Behera

Abstract Quantum Natural Language Processing (QNLP) stands at the forefront of interdisciplinary research, amalgamating quantum computing principles with natural language processing techniques to revolutionize sentiment analysis (SA). Departing from traditional methods grounded in classical computation, QNLP holds promise for surmounting specific challenges within SA, offering a novel lens through which to discern the intricacies of human emotions, attitudes, and assessments in textual data. A burgeoning body of research underscores the efficacy of quantum NLP in augmenting SA performance, underscoring its potential to surpass classical approaches in capturing the nuances of human sentiment. In the contemporary landscape, quantum cognitive-inspired models have emerged as formidable tools for SA, demonstrating superior efficacy compared to their classical counterparts. This chapter embarks on a rigorous exploration of QNLP, commencing with a meticulous examination of quantum probability and cognition. By elucidating the advantages afforded by these quantum concepts over classical cognitive-based SA theories, we lay the groundwork for understanding the transformative potential of QNLP. Subsequently, we delve into various approaches devised to tackle the inherent challenges of SA tasks, showcasing how quantum techniques can furnish innovative solutions. From quantum-enhanced feature extraction to sophisticated sentiment classification algorithms, we scrutinize the evolving landscape of SA methodologies enriched by quantum principles. Moreover, this chapter critically evaluates the current state of QNLP research, acknowledging both its achievements and limitations. While quantum-inspired models have

M. Kayal · M. Sahoo · J. Pati · R. K. Behera (✉)
Department of Computer Science and Engineering, Indian Institute of Information Technology Ranchi, Ranchi, Jharkhand, India
e-mail: ranjan@iiitranchi.ac.in

M. Kayal
e-mail: mrinmoy02.rs21@iiitranchi.ac.in

M. Sahoo
e-mail: mohinikanta01.rs21@iiitranchi.ac.in

J. Pati
e-mail: jayadeeppati@iiitranchi.ac.in

S. Dehuri et al. (eds.), *Advances in Quantum Inspired Artificial Intelligence*, Intelligent Systems Reference Library 274, https://doi.org/10.1007/978-3-031-89905-8_8

shown remarkable promise, practical constraints and theoretical boundaries necessitate further investigation to unlock their full potential. By illuminating these nuances, we aim to foster a deeper understanding of QNLP's complexities and inspire future research endeavors aimed at refining SA techniques for greater precision and insight.

Keywords Quantum NLP · Sentiment analysis · Cognitive probability · Machine learning

8.1 Introduction

Quantum Natural Language Processing (QNLP) is an exciting new approach that combines the power of quantum computing with the ability to understand and process human language. It brings together advanced concepts from both fields to explore new ways of handling language-related tasks [1]. Its potential in sentiment analysis stems from quantum computing's unique ability to handle complex, ambiguous, and high-dimensional data. Natural language, with its inherent ambiguity and layered meanings, benefits from QNLP's capacity to capture subtle linguistic nuances. Quantum systems excel in representing data in high-dimensional spaces, a critical requirement in NLP where words, phrases, and sentences are mapped to such spaces. Quantum states and operations enable efficient encoding and manipulation of these vectors, enhancing the representation and analysis of linguistic data [2].

Quantum computing's massive parallelism allows for the simultaneous evaluation of multiple solutions, significantly accelerating sentiment analysis tasks that often involve processing vast datasets. Quantum entanglement [3] offers a novel way to model dependencies between words in a sentence, potentially improving the accuracy of sentiment analysis models. Algorithms like Grover's search and quantum neural networks introduce methodologies uniquely suited to NLP, enabling solutions beyond the reach of classical computing. As quantum technology advances, it is expected to surpass classical systems in specific tasks, highlighting the importance of investing in QNLP research to prepare for a future where quantum computing becomes integral to sentiment analysis and NLP.

Sentiment analysis, also known as opinion mining, identifies and categorizes sentiments in text as positive, negative, or neutral. It evaluates emotions, attitudes, and opinions toward entities such as products, services, or events. NLP, as a branch of artificial intelligence, facilitates tasks like translation, categorization, and sentiment analysis. Applications span diverse domains, including business intelligence (e.g., customer feedback, brand monitoring), market research (e.g., trend analysis, competitor analysis), social media monitoring, customer service, human-computer interaction, and healthcare systems. Sentiment analysis operates at three levels-document level, sentence level, and aspect level-and is often used to analyze public opinion derived from structured discussions.

Aspect-Based Sentiment Analysis (ABSA) is a detailed and nuanced method that digs deeper than just determining whether a text is positive, negative, or neutral.

Instead, it focuses on pinpointing sentiments tied to specific features or elements of a product, service, or topic being discussed. For instance, take the sentence: "The hotel had stunning views and a cozy atmosphere, but the room was way too small." Here, ABSA would break down the sentiment for each aspect: "views" (positive), "atmosphere" (positive), and "room size" (negative). This allows for a more precise understanding of what people like or dislike, making it incredibly useful for businesses or researchers who want to improve specific areas rather than making broad generalizations. ABSA consists of two key components: aspect extraction and sentiment classification. It employs techniques such as rule-based, machine learning, and hybrid approaches. Applications of ABSA include customer feedback analysis, product reviews, service industry optimization (e.g., hospitality, healthcare), social media monitoring, market research, and financial investment analysis.

By leveraging quantum computing's advanced data processing capabilities, NLP tasks can be optimized, and machine learning models enhanced, revolutionizing ABSA. Quantum computing's integration into ABSA promises more accurate, efficient, and scalable sentiment analysis solutions, enabling deeper and more actionable insights from textual data. As quantum technology evolves, its impact on ABSA and other NLP fields is expected to grow, offering unprecedented opportunities for innovation.

Sentiment analysis focuses on identifying positive or negative opinions, while emotion detection identifies specific emotions such as happiness, anger, or fear. Sentiment analysis involves polarity and subjectivity measures, which help evaluate user opinions in reviews, social media, and other textual data. It employs two primary approaches: lexicon-based (unsupervised, using predefined lexicons) and machine learning-based (supervised, relying on feature extraction and model training). Applications of sentiment analysis span movies, sports, politics, market analysis, and service industries, making it a vital tool for understanding user opinions and feedback.

The contributions of this chapter are outlined as follows:

(1) The application of quantum theory to aspect-based sentiment analysis (SA) is introduced.
(2) A comprehensive review of quantum-inspired aspect-based SA using natural language processing (NLP) is presented.
(3) A comparative analysis between quantum classifier-based ABSA and classical classifier-based ABSA is conducted.
(4) Finally, the challenges associated with quantum ABSA and the limitations of quantum techniques are discussed.

The subsequent organization of the chapter is as follows: In the second section, we present some of the existing related work based on quantum computation on ABSA. The the fundamental to quantum theory and its applications in Aspect based sentiment analysis is discussed in Sect. 8.3. The methodology involve in quantum inspired classifier for ABSA is presented in Sect. 8.4 We have also discussed the difference between traditional and Quantum classifier for aspect based sentiment prediction in this section. The issues and challenges exist in adopting quantum theory in sentiment analysis is presented in Sect. 8.5. The research opportunities in this area is presented in Sect. 8.6. The last section discusses the conclusion and future direction of the work presented in this chapter.

8.2 Related Work

With the growing number of electronic documents on websites and the rapid expansion of social media platforms, sentiment analysis has gained immense significance. Millions of users express their opinions through comments and reviews, which are valuable for understanding public sentiment. Various research works have proposed different techniques for sentiment and emotion analysis including machine learning and deep learning models.

Deep learning approaches have been leveraged for sentiment and emotion analysis. Chiorrini et al. [4] utilized BERT for sentiment analysis of Twitter data, achieving an accuracy of 0.92 for emotion classification and 0.90 for sentiment classification. Feature selection and classification techniques have been explored in multiple studies. The application of sentiment analysis to regional languages has also been explored. Prasad et al. [5] implemented sentiment classification techniques for Indian language tweets using decision trees. Alsaeedi and Khan [6] provided a comprehensive study on sentiment analysis methodologies applied to Twitter data, highlighting various techniques used in this domain.

A combined approach to sentiment analysis was introduced by Prabowo and Thelwall [7], where multiple classifiers contributed to improving overall accuracy. Lexicon-based approaches have also been utilized; Nausheen and Begum [8] employed a lexicon-based technique to analyze political sentiment and predict election outcomes based on Twitter data. Piotrowski et al. [9] investigated the application of quantum-like probabilistic models in text classification, particularly focusing on sentiment analysis. Their study demonstrated that quantum interference effects could enhance sentiment prediction by capturing contextual dependencies that classical models often overlook. The research provided a novel perspective on utilizing quantum mechanics principles in natural language processing (NLP). Ying et al. [10] introduced a quantum-inspired tensor network approach for sentiment analysis, which leveraged quantum-inspired structures to better capture complex word dependencies. Their results indicated that tensor networks could effectively model intricate relationships in text data, leading to improved classification accuracy compared to conventional methods. Zhang et al. [11] proposed a hybrid quantum-classical framework for sentiment classification, where quantum circuits were employed for feature transformation. By integrating quantum computation into traditional machine learning pipelines, the framework demonstrated enhanced capability in processing high-dimensional textual data, showcasing the potential of quantum computing in NLP tasks.

Li et al. [12] explored the feasibility of quantum-inspired variational circuits for sentiment analysis, emphasizing the advantages of quantum-enhanced representations. Their study highlighted how variational circuits could encode complex linguistic features, leading to improved performance in sentiment classification tasks. The research contributed to the growing body of work on quantum-inspired deep learning techniques. Melucci et al. [13] applied quantum-inspired similarity measures in sentiment analysis, demonstrating their effectiveness in improving classification accuracy on benchmark datasets. Their approach leveraged quantum principles to refine text similarity computation, addressing limitations in traditional distance-based methods. The study provided empirical evidence supporting the applicability of quantum-inspired metrics in NLP. Wang et al. [14] investigated the use of quantum-inspired neural networks for sentiment classification, showcasing their advantages over conventional deep learning models. By incorporating quantum principles into neural architectures, their approach enhanced sentiment classification performance, suggesting that quantum-inspired methods could offer new directions for improving text analysis tasks.

Sentiment analysis has gained significant importance across various domains, including politics, finance, mental health, and regional languages, with machine learning (ML) and deep learning (DL) techniques achieving notable accuracy. Methods such as text preprocessing, supervised learning models, and advanced frameworks like BERT and hybrid approaches have been widely used. Recent advancements in quantum-inspired models, leveraging principles like quantum interference, tensor networks, and variational circuits, have shown promise in enhancing sentiment analysis by capturing complex contextual dependencies and improving classification accuracy. However, the integration of quantum computing into sentiment analysis, particularly in Aspect-Based Sentiment Analysis (ABSA), remains underexplored. This presents a critical research gap, as quantum computing holds the potential to process high-dimensional data and model intricate linguistic relationships more effectively. Further research is needed to bridge classical and quantum approaches, enabling more efficient and innovative solutions for sentiment analysis in the face of growing textual data complexity.

8.3 Background Details

Quantum computing has the potential to revolutionize Aspect-Based Sentiment Analysis (ABSA) by addressing the scalability and efficiency limitations of classical computing. Traditional ABSA methods struggle with handling large datasets, recognizing complex patterns, and accurately linking sentiment with specific aspects due to computational constraints. Quantum computing leverages principles such as superposition and entanglement to perform parallel computations.

Quantum Machine Learning (QML) models leverage the unique properties of quantum states to store and process information in ways that are far more efficient than traditional methods. This approach enables these models to identify and analyze patterns with greater precision, particularly when it comes to understanding and interpreting emotions in sentiment analysis tasks. Quantum Natural Language Processing (QNLP) further improves context understanding by capturing intricate dependencies between aspect terms and sentiment expressions, leading to more nuanced sentiment classification. Additionally, quantum algorithms such as Quantum Support Vector Machines (QSVM) and Variational Quantum Circuits offer superior classification capabilities with higher accuracy and efficiency. Quantum-enhanced transformers have the potential to accelerate training and inference in models like BERT, reducing computational overhead while maintaining performance. As quantum technology advances, it promises to significantly enhance the speed, accuracy, and scalability of ABSA, making it a powerful tool for sentiment analysis in real-world applications.

Some of the background details related to aspect-based sentiment analysis and quantum theory are presented below.

8.3.1 Aspect Based Sentiment Analysis (ABSA)

Aspect-Based Sentiment Analysis (ABSA) is a fine-grained sentiment analysis technique that identifies sentiments expressed towards specific aspects of an entity, rather than analyzing the overall sentiment in a piece of text. Traditional sentiment analysis focuses on determining whether a review, comment, or opinion is positive, negative, or neutral. However, ABSA breaks down the sentiment analysis process into multiple aspects and determines sentiments associated with each.

Considering a review that says, "The sound quality of these headphones is fantastic, but the build feels cheap." A basic sentiment analysis system might just see this as a neutral statement since it has both positive and negative opinions. However, an Aspect-Based Sentiment Analysis (ABSA) system would break it down further-recognizing "sound quality" as a positive aspect and "build" as a negative one. This way, ABSA provides a more detailed understanding of the review.

Key Components of Aspect-Based Sentiment Analysis

ABSA typically consists of multiple sub-tasks, including:

1. Aspect Term Extraction (ATE) [15]: Aspect Term Extraction involves identifying specific words or phrases in a text that represent aspects of an entity. This is crucial because aspects define what features or components of a product or service are being discussed. In the sentence "The hotel room was spacious, but the Wi-Fi was unreliable.", "room" and "Wi-Fi" are aspect terms. ATE employs techniques such as Named Entity Recognition (NER), dependency parsing, and deep learning-based sequence tagging to extract these aspect terms effectively.
2. Aspect Category Detection (ACD) [16]: Sometimes, aspects are not explicitly mentioned in the text but are implied. Aspect Category Detection (ACD) is the process of classifying a given text into predefined aspect categories. For instance, a review may mention "The waiter was rude", which does not explicitly state

"customer service" but falls under that category. Common categories in different domains include "food quality", "customer service", "ambience", and "pricing" in restaurant reviews. ACD is typically done using supervised classification models, topic modeling, and neural networks trained on labeled datasets.

3. Sentiment Polarity Detection [17]: Sentiment Polarity Detection is the process of determining whether the sentiment expressed towards an aspect is positive, negative, or neutral. This is the core task of ABSA, as it assigns a sentiment score to each detected aspect. For example, in "The battery life is fantastic", but the design is outdated., fantastic indicates a positive sentiment towards battery life, while "outdated" conveys a negative sentiment towards the design. This task is typically performed using lexicon-based methods, rule-based approaches, or deep learning models such as Long Short-Term Memory (LSTM) networks and transformer-based architectures like BERT.
4. Opinion Term Extraction (OTE) [18]: Opinion Term Extraction (OTE) involves identifying the specific words or phrases in a text that express sentiment. These words play a crucial role in determining the polarity of an aspect. In the sentence "The laptop's display is stunning, but its keyboard feels cheap.", "stunning" and "cheap" are opinion terms that respectively indicate positive and negative sentiment towards the "display" and "keyboard" aspects. OTE is often performed using sequence labeling techniques such as Conditional Random Fields (CRF) or deep learning methods like BiLSTM-CRF models.
5. Aspect-Sentiment Pairing [19]: Aspect-Sentiment Pairing is the process of linking the detected aspects with their corresponding sentiment expressions. This ensures that each aspect is correctly associated with its sentiment. For instance, in "The processor is powerful, but the battery drains quickly.", the aspect "processor" should be linked to "powerful" (positive sentiment), and "battery" should be linked to "drains quickly" (negative sentiment). Aspect-Sentiment Pairing is often achieved using attention mechanisms in deep learning models or dependency parsing techniques.

8.3.2 Fundamental of Quantum Theory

Quantum theory serves as a foundational framework in physics, describing how matter and energy behave at microscopic scales, including atoms and subatomic particles. At its core are quantum states, which characterize a quantum system and are mathematically expressed as vectors in a complex Hilbert space. Unlike classical mechanics, quantum mechanics introduces pioneering concepts such as superposition, entanglement, and wave-particle duality, challenging traditional views of reality and driving advancements in areas like computing, cryptography, and materials science.

8.3.3 Hilbert Space and State Vectors

A quantum state can be thought of as a state vector $|\psi\rangle$ that lives in a special kind of mathematical space called a Hilbert space $\mathcal{H}$. This space is like a well-organized playground for vectors, where we can measure their sizes (norms) and the angles between them using an inner product.

State Vector:

$$|\psi\rangle \in \mathcal{H}$$

Inner Product: For two state vectors $|\psi\rangle$ and $|\phi\rangle$, the inner product is denoted as $\langle\psi|\phi\rangle$.

8.3.4 Basis States and Superposition

In quantum mechanics, a Hilbert space is often described using a set of special vectors called orthonormal basis vectors, which we can represent as $\{|i\rangle\}$. These vectors are like the building blocks of the space, and they have a unique property: they are perpendicular to each other and normalized, meaning each vector points in a completely independent direction and has a length of one. Mathematically, this is expressed as $\langle i, j\rangle = \delta_{ij}$, where δ_{ij} is the Kronecker delta that represents the inner product and it is zero when applied for different vector, while the inner product of a vector with itself is one.

8.3.5 Quantum Superposition

Quantum superposition allows a quantum state to be in a combination of multiple basis states simultaneously.

Example: For a single qubit, the basis states are $|0\rangle$ and $|1\rangle$.

$$|\psi\rangle = \alpha|0\rangle + \beta|1\rangle$$

where α and β are complex numbers such that $|\alpha|^2 + |\beta|^2 = 1$.

8.3.6 Dirac Notation

Quantum states are often written in Dirac (bra-ket) notation.

Ket: $|\psi\rangle$ represents a column vector.

Bra: $\langle\psi|$ represents the conjugate transpose of the ket, a row vector.

For a qubit:

$$|\psi\rangle = \begin{pmatrix} \alpha \\ \beta \end{pmatrix}$$

where α and β are complex numbers.

8.3.7 Measurement and Probability

The probability of measuring a quantum state $|\psi\rangle$ in a basis state $|i\rangle$ is given by the square of the inner product's magnitude:

$$P(i) = |\langle i|\psi\rangle|^2$$

8.3.8 Quantum Entanglement

Entanglement is a property where the quantum state of one particle cannot be described independently of the state of another.

Two-Qubit Entangled State (Bell State):

$$|\Phi^+\rangle = \frac{1}{\sqrt{2}}(|00\rangle + |11\rangle)$$

Here, the state of each qubit is dependent on the other, exhibiting correlations that are stronger than classical systems.

8.4 Quantum-Inspired Aspect-Based Sentiment Analysis (QIASA)

Quantum-Inspired Aspect-Based Sentiment Analysis (QIASA) is an emerging methodology which integrates principles of quantum computing with classical sentiment analysis to improve efficiency, accuracy, and scalability. QIASA specifically focuses on extracting and analyzing sentiments towards specific aspects within a text, leveraging the unique computational advantages of quantum computing.

Traditional aspect-based sentiment analysis (ABSA) relies on classical machine learning (ML) models that process textual data using conventional feature extraction, classification, and deep learning-based natural language processing (NLP) techniques. However, these methods often struggle with computational complexity, feature representation limitations, and optimization inefficiencies when dealing with large-scale sentiment datasets. QIASA addresses these challenges by incorporating

Table 8.1 A comprehensive analysis on quantum classifiers and its applications

Algorithm	Classification	Quantum	Domains	Advantages	Limitations
HQC	Binary	Yes	Appendicitis	Enhances feature representation via quantum coherence	High computational time; difficulty with complex data
QIBC	Binary	Yes	Image, Text	Captures intricate data patterns using quantum entanglement	Requires too many features, increasing complexity
QNMC	Binary and Multi-class	Yes	Ionosphere, Cancer Dataset Diabetes	Achieves high accuracy through optimized quantum kernels	Ineffective for cancer data
VQC	Multi-level Classification	Yes	Wine dataset Brest cancer dataset	Adaptive learning through quantum circuit optimization	High computational cost
QSVM	Multi-level Classification	Yes	Wine dataset Cancer dataset	Robust classification via quantum support vectors	Difficult to implement
APM	Multi-level Classification	No	Wine dataset Synthetic dataset	Efficient on classical computers, offering reliable results	Limited to classical systems

quantum-inspired and hybrid quantum-classical approaches, offering enhanced computational speed, better feature encoding, and improved classification accuracy. A comparative analysis on quantum classifiers along with its advantages and limitations is presented in Table 8.1.

In this chapter we have made a comparative analysis of quantum classifiers and classical ML classifiers in sentiment classification tasks. Figures 8.1and 8.2 illustrate the methodologies applied in this comparison. The following techniques can be used to enhance the performance of QIASA:

- Quantum ML (QML)
- Quantum-Inspired ML (QIML)
- Hybrid Quantum-Classical ML.

8.4.1 Quantum Machine Learning (QML)

QML leverages quantum computing principles to process and analyze sentiment data at a significantly enhanced computational speed. Dennis et al. [20] implemented quantum annealer devices using SVM, known as QA-SVM. Their study applied quantum annealing to train and optimize SVM models using Quadratic Unconstrained Binary Optimization (QUBO) equations, minimizing cost energy effectively. By leveraging the properties of quantum annealing, they improved final model accuracy through quantum feature selection. Rebentrost et al. [21] extended this approach by applying quantum support vector machines (QSVMs), utilizing a non-sparse matrix representation to efficiently classify sentiment data. This method showed potential for scalability with large datasets. Further, Silva et al. [22] introduced a novel QNN model called the Quantum Perceptron Over Field, demonstrating quantum-enhanced learning capabilities in sentiment analysis tasks. Other research, such as [23], explored linear regression models implemented on quantum computers, paving the way for quantum-based sentiment prediction models with increased efficiency over classical counterparts.

8.4.2 Quantum-Inspired Machine Learning (QIML)

QIML applies quantum computing principles to improve the performance of classical machine learning models. QIML leverages concepts such as quantum probability theory, quantum state representation, and superposition to expand decision-making capabilities beyond classical methods.

A study by [24] introduced a quantum-inspired binary classifier integrating K-Nearest Neighbors (KNN) and SVM. This classifier utilized principles of quantum decision theory and quantum probability to enhance precision, recall, and F1-score in sentiment classification tasks. By applying quantum superposition, the classifier expanded the decision space, improving generalization over classical models. Sergoli et al. [25] proposed the Helstrom quantum centroid (HQC) classifier, a quantum-inspired method for binary classification based on quantum centroid measurement. Their model significantly improved sentiment classification accuracy by leveraging quantum probability distributions. In [26], the same authors introduced the quantum nearest mean classifier (QNMC), which applied a three-step quantum encoding process:

- Encoding: Mapping classical data points to quantum representations through density pattern representation.
- Distance Calculation: Computing distances among quantum density patterns to identify sentiment clusters.
- Decoding: Converting quantum results into classified sentiment data.

These quantum-inspired methods have demonstrated higher accuracy and better decision boundary adaptability compared to classical ML classifiers, making them promising candidates for ABSA tasks.

8.4.3 Hybrid Quantum-Classical Machine Learning

It combines quantum computing techniques with classical machine learning to optimize performance while reducing computational cost. This approach is particularly useful for QIASA, as it balances the advantages of quantum speedup with the practicality of classical processing.

Soumik et al. [27] proposed a Variational Quantum Classifier (VQC), which mapped input features onto a quantum system and optimized parameters using a hybrid variational algorithm. Their approach achieved high precision with minimal training parameters, demonstrating quantum advantage in sentiment analysis. Another study [28] introduced novel quantum algorithms for enhancing classical ML techniques, running exclusively on quantum simulators. While these methods improved model accuracy, their reliance on simulators posed a limitation, as real quantum hardware implementation remains a challenge due to noise and gate errors.

8.4.3.1 Quantum Classifier Versus Classical Classifier

A quantum classifier is a type of machine learning algorithm built specifically for quantum computers. It takes advantage of quantum mechanics concepts like superposition, entanglement, and interference to classify data in ways that classical computers might struggle with. Quantum classifiers are particularly promising for solving complex problems that are computationally expensive for classical systems. The workflow of a quantum classifier typically consists of seven phases: data collection, pre-processing, encoding, feature extraction, training and evaluation, testing, and decoding. This process is illustrated in Fig. 8.1.

In contrast, a classical classifier follows a more straightforward pipeline, as shown in Fig. 8.2, which does not require encoding and decoding steps. Classical classifiers operate directly on classical data, while quantum classifiers require the transformation of classical data into quantum states (quantum data) through encoding and decoding processes.

The difference between classical, quantum and hybrid classifier is presented in Table 8.2. Some of the important difference between classical and quantum classifiers are described through the following parameters such as :

- Encoding and Decoding: Quantum classifiers require encoding to map classical data (e.g., feature vectors (X) into quantum states $|\Psi\rangle$ and decoding to transform quantum results back into classical form. Classical classifiers do not require encoding or decoding, as they operate directly on classical data.

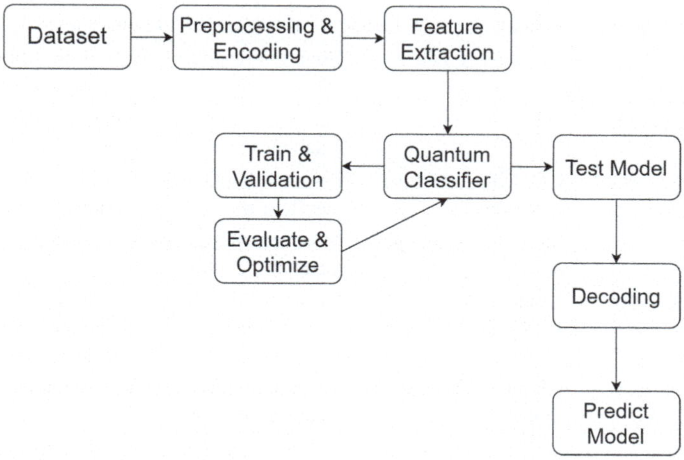

Fig. 8.1 Pipeline of quantum classifier

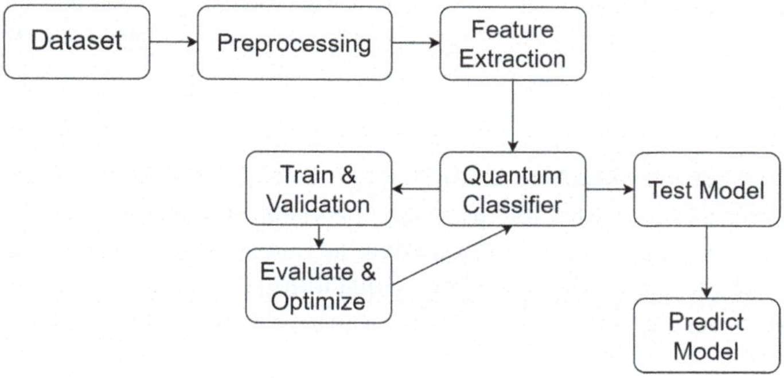

Fig. 8.2 Pipeline of classical classifier

- Data Compatibility: Quantum classifiers can work with both quantum data and classical data (via encoding). However, some quantum-inspired algorithms are designed specifically for classical data. Classical classifiers are limited to classical data.
- Pre-processing: Both quantum and classical classifiers use pre-processing techniques, such as handling missing values, feature reduction, and dataset splitting (train, test, validation). However, quantum pre-processing may involve additional steps to ensure compatibility with quantum hardware (e.g., reducing features to match the number of available qubits).
- Training and Evaluation: Quantum classifiers use quantum circuits or hybrid quantum-classical algorithms to train the model and minimize the cost function. Classical classifiers rely on classical optimization techniques (e.g., gradient descent) for training.

Table 8.2 Comparison of classical, quantum, and hybrid classifiers in sentiment analysis

Feature	Classical classifier	Quantum-inspired classifier	Hybrid quantum-classical classifier
Encoding required?	No	Yes (Classical to Quantum)	Partial (Hybrid Mapping)
Computational speed	Limited by classical hardware	Faster with quantum parallelism	Balanced (Leverages both)
Scalability	Struggles with large datasets	More scalable due to quantum parallelism	Moderate scalability
Optimization Techniques	Gradient descent, Backpropagation	Quantum annealing, QAOA, VQAs	Hybrid quantum-classical optimization
Training cost	High for deep learning models	Potentially lower with quantum speedup	Moderate
Interpretability	Well-understood ML models	Complex quantum state representations	Hybrid methods improve interpretability
Hardware dependency	Classical GPUs and TPUs	Requires quantum processors	Uses both quantum and classical hardware

- Testing and Performance Metrics: Both quantum and classical classifiers evaluate performance using metrics like accuracy, precision, recall, and F1-score. Quantum classifiers may also measure quantum advantage in terms of computational speed or resource efficiency. Testing in quantum classifiers involves transforming quantum results into classical form for interpretation (e.g., using a confusion matrix).

8.5 Issues and Challenges

The integration of quantum computing concepts into Aspect-Based Sentiment Analysis (ABSA) holds the promise of enhancing sentiment analysis capabilities by leveraging the unique properties of quantum mechanics. When the quantum model runs on the quantum computer but it does not run their classical classifier such as Shor's algorithm [29], Grover's algorithms [30]. Quantum ML is a new trends in our research field to solve the optimization problems, and it has to improve the performance.

8.5.1 Challenges and Solutions

Quantum-inspired computing has shown great potential in natural language processing (NLP) and aspect-based sentiment analysis (ABSA). However, several challenges hinder its practical application. One of the primary issues is computational complexity and resource constraints. Quantum-inspired algorithms, especially those utilizing quantum entanglement and superposition, demand significant computational power. Although classical computers can simulate quantum principles, this requires high-performance computing resources, making large-scale implementation difficult. Additionally, integrating quantum-inspired models into existing NLP frameworks presents challenges, as current architectures are designed primarily for classical machine learning and deep learning methods, necessitating modifications to accommodate quantum-enhanced processing. A comparative analysis of various quantum algorithms suitable for various applications is presented in Table 8.3.

Another major challenge is the limited scalability of small-scale quantum computers. While companies like IBM, Rigetti, Xanadu, Q-Wave, Google, and Microsoft have developed quantum processors, the number of available qubits remains insufficient for large-scale ABSA tasks. Researchers have developed noisy quantum systems and smaller-scale quantum-inspired models, but these lack the capability to handle extensive datasets effectively. Furthermore, classical machine learning techniques used in sentiment analysis involve large feature sets, which are difficult to process on current quantum devices due to their limited computational capacity.

Encoding methods pose another significant hurdle in quantum-inspired ABSA. The conversion of classical data into quantum states is highly complex and energy-intensive. Researchers have attempted to design improved encoding techniques, such as binary quantum classifiers, to handle noisy quantum states more effectively. However, selecting the optimal encoding method remains a challenge, as quantum data representation differs significantly from classical data structures. Additionally, feature representation in quantum systems, such as amplitude encoding and Hilbert space embeddings, requires novel approaches to maintain sentiment nuances while transforming textual data into quantum-compatible formats.

Interpretability and explainability of quantum-inspired sentiment models also present difficulties. These models operate in high-dimensional spaces, making their decision-making processes harder to understand compared to traditional deep learning approaches. Classical models have established interpretability frameworks, but similar techniques are still in development for quantum-inspired models. Furthermore, ethical considerations and bias mitigation remain critical concerns. Quantum-inspired models, like classical NLP models, can inherit biases from training datasets. Ensuring fairness, minimizing biases, and improving ethical AI principles in quantum-inspired sentiment analysis is essential for practical deployment.

Scalability and stability of quantum-inspired models also require significant improvements. Current implementations are tested on small datasets, and scaling them to analyze large-scale sentiment data, such as millions of social media posts, remains a challenge. Additionally, quantum computing systems suffer from noise

Table 8.3 A comparative analysis of various problem solve by Quantum algorithms and their role in sentiment analysis

Problem	Algorithm	Application in sentiment analysis	Speed up	Advantages
Factorization	Shor's Algorithm [31]	Encryption-breaking for secure NLP systems	Exponential	Enhances data security in sentiment analysis pipelines
Searching in unstructured data	Grover's Algorithm [32]	Aspect extraction and sentiment classification	Quadratic	Faster search for relevant features in large datasets
Linear systems of equations	HHL Algorithm [33]	Dimensionality reduction in NLP models	Exponential	Efficiently processes high-dimensional textual data
QUBO (Quadratic Unconstrained Binary Optimization)	D-Wave quantum computer [34]	Optimization of sentiment classification models	NP-hard	Solves complex optimization problems in ABSA
Quantum Neural Networks (QNN)	Quantum-inspired neural networks [35]	Sentiment and emotion classification	Polynomial	Captures complex linguistic patterns with quantum parallelism
Quantum Tensor Networks	Tensor network models [36]	Sentiment analysis of multi-language datasets	Polynomial	Handles high-dimensional data and cross-lingual sentiment analysis
Quantum Variational Circuits	Variational quantum classifiers [37]	Fine-grained sentiment analysis (e.g., ABSA)	Polynomial	Adaptable to small-scale quantum devices for practical NLP tasks
Quantum Interference Models	Quantum-inspired probabilistic models [38]	Contextual sentiment analysis	Polynomial	Captures subtle dependencies in text for improved accuracy

due to quantum decoherence, which can lead to instability in sentiment predictions. Addressing these challenges is crucial for making quantum-inspired ABSA a viable alternative to classical sentiment analysis models.

8.6 Research Opportunities

Despite these challenges, quantum-inspired techniques offer several promising research opportunities. By addressing computational limitations and improving integration with classical models, researchers can enhance the effectiveness of ABSA. Below are some key areas of research that can drive progress in quantum-inspired sentiment analysis:

- Hybrid Quantum-Classical Approaches: Combining quantum-inspired techniques with classical deep learning models can enhance computational efficiency and improve sentiment classification accuracy. Using quantum embeddings alongside transformer architectures can offer better contextual representation in ABSA.
- Quantum-Inspired Sentiment Representation: Developing novel methods to encode sentiment polarity using quantum probability distributions can improve sentiment classification. Leveraging quantum entanglement in word and sentence embeddings can help capture intricate sentiment relationships more effectively.
- Advanced Quantum Optimization and Reinforcement Learning: Integrating quantum-inspired reinforcement learning techniques, such as quantum Markov chains and quantum Bellman equations, can optimize aspect extraction and sentiment classification for better predictive performance.
- Development of Quantum-Friendly Datasets and Interpretability Tools: Creating benchmark datasets specifically designed for quantum-inspired NLP models can improve evaluation and standardization. Additionally, developing interpretability frameworks for quantum-enhanced models will make their decision-making processes more transparent and explainable.
- Ethical AI and Bias Mitigation in Quantum Models: Implementing quantum fairness constraints and bias-aware embeddings can enhance fairness and reduce biases in sentiment classification. Ensuring equitable sentiment analysis across different demographic groups is essential for responsible AI development.
- Practical Deployment on Noisy Intermediate-Scale Quantum (NISQ) Devices: Testing quantum-inspired ABSA in real-world scenarios on today's NISQ devices can help turn theoretical breakthroughs into practical solutions. This will help researchers test quantum models in practical sentiment analysis tasks and refine their performance.

8.7 Conclusion

Sentiment analysis is necessary for different areas such as medicine, product review, business, etc. in our daily life. In this paper, we proposed to analyze Quantum-inspired Aspect-Based Sentiment Analysis (QIASA) using NLP. QIASA represented a promising frontier in Natural Language Processing (NLP), blending principles from quantum computing with traditional sentiment analysis techniques. QIASA held the potential to enhance sentiment analysis by leveraging quantum principles like superposition and entanglement. These principles could theoretically capture more nuanced relationships between aspects and sentiments, leading to more accurate analyses. Despite its promise, QIASA faced significant challenges. Implementing quantum-inspired algorithms in practical applications. QIASA is complex as advanced hardware enabling quantum processing and sophisticated mathematical frameworks are required. While quantum-inspired approaches offer new avenues, they are not meant to replace traditional NLP techniques but rather complement them. Integration with existing methods remains crucial for practical implementation and real-world applications. Research in QIASA is still in its early stages. More studies are needed to validate its effectiveness across different domains and to develop scalable algorithms that can handle real-world data efficiently

In the future, we use the big and complex dataset with an increase in the number of labels for quantum inspired ABSA. For better predictions, we apply the different types of deep learning techniques classifier to predict the emotion states, and product reviews of the people using the Indian language dataset, product dataset, etc.

References

1. Guarasci, R., De Pietro, G., Esposito, M.: Quantum natural language processing: challenges and opportunities. Appl. Sci. **12**, 5651 (2022)
2. Wu, S., Li, J., Zhang, P., Zhang, Y.: natural language processing meets quantum physics: a survey and categorization, pp. 3172–3182 (2021)
3. Życzkowski, K., Horodecki, P., Horodecki, M., Horodecki, R.: Dynamics of quantum entanglement. Phys. Rev. A **65**, 012101 (2001)
4. Chiorrini, A., Diamantini, C., Mircoli, A., Potena, D.: Emotion and sentiment analysis of tweets using bert, vol. 3 (2021)
5. Prasad, S.S., Kumar, J., Prabhakar, D.K., Pal, S.: sentiment classification: an approach for Indian language tweets using decision tree, pp. 656–663. Springer, Berlin (2015)
6. Alsaeedi, A., Khan, M.Z.: A study on sentiment analysis techniques of twitter data. Int. J. Adv. Comput. Sci. Appl. **10**, 361–374 (2019)
7. Prabowo, R., Thelwall, M.: Sentiment analysis: a combined approach. J. Informetrics **3**, 143–157 (2009)
8. Nausheen, F., Begum, S.H.: sentiment analysis to predict election results using python, pp. 1259–1262. IEEE (2018)
9. Piotrowski, E.W., Sladkowski, J.: Quantum-like model of subjective expected utility. Int. J. Theor. Phys. **42**, 1101–1110 (2003)
10. Ying, M., et al.: Quantum tensor networks for sentiment analysis. Quantum Inf. Process. **20**, 1–15 (2021)

11. Zhang, R., et al.: Hybrid quantum-classical sentiment analysis using variational circuits. Quantum Mach. Learn. J. **4**, 1–12 (2022)
12. Li, C., et al.: Quantum variational circuits for natural language processing. IEEE Trans. Quantum Eng. **4**, 1–10 (2023)
13. Melucci, M.: Quantum-inspired similarity measures for sentiment classification. Inf. Process. Manag. **57**, 102240 (2020)
14. Wang, H., et al.: Quantum-inspired neural networks for sentiment analysis. J. Artif. Intell. Res. **76**, 1–20 (2023)
15. Kumari, A., Sahoo, S.P., Behera, R.K., Sahoo, B.: Supervised machine learning for link prediction using path-based similarity features, pp. 1–7. IEEE (2020)
16. Chebolu, S.U.S., Rosso, P., Kar, S., Solorio, T.: Survey on aspect category detection. ACM Comput. Surv. **55**, 1–37 (2022)
17. Calefato, F., Lanubile, F., Maiorano, F., Novielli, N.: Sentiment polarity detection for software development, pp. 128–128 (2018)
18. Kumari, A., Sahoo, B., Behera, R.K.: Mitigating cold-start delay using warm-start containers in serverless platform, pp. 1–6. IEEE (2022)
19. Nazir, A., Rao, Y.: Iaotp: An interactive end-to-end solution for aspect-opinion term pairs extraction, pp. 1588–1598 (2022)
20. Willsch, D., Willsch, M., De Raedt, H., Michielsen, K.: Support vector machines on the d-wave quantum annealer. Comput. Phys. Commun. **248**, 107006 (2020)
21. Rebentrost, P., Mohseni, M., Lloyd, S.: Quantum support vector machine for big data classification. Phys. Rev. Lett. **113**, 130503 (2014)
22. da Silva, A.J., Ludermir, T.B., de Oliveira, W.R.: Quantum perceptron over a field and neural network architecture selection in a quantum computer. Neural Netw. **76**, 55–64 (2016)
23. Schuld, M., Sinayskiy, I., Petruccione, F.: Prediction by linear regression on a quantum computer. Phys. Rev. A **94**, 022342 (2016)
24. Tiwari, P., Melucci, M.: Towards a quantum-inspired binary classifier. IEEE Access **7**, 42354–42372 (2019)
25. Sergioli, G., Giuntini, R., Freytes, H.: A new quantum approach to binary classification. PloS one **14**, e0216224 (2019)
26. Sergioli, G., et al.: Quantum-inspired minimum distance classification in a biomedical context. Int. J. Quantum Inf. **16**, 1840011 (2018)
27. Adhikary, S., Dangwal, S., Bhowmik, D.: Supervised learning with a quantum classifier using multi-level systems. Quantum Inf. Process. **19**, 1–12 (2020)
28. Chakraborty, S., Shaikh, S.H., Chakrabarti, A., Ghosh, R.: A hybrid quantum feature selection algorithm using a quantum inspired graph theoretic approach. Appl. Intell. **50**, 1775–1793 (2020)
29. Shor, P.W.: Polynomial-time algorithms for prime factorization and discrete logarithms on a quantum computer. SIAM Rev. **41**, 303–332 (1999)
30. Grover, L.K.: A fast quantum mechanical algorithm for database search, pp. 212–219 (1996)
31. Ugwuishiwu, C., Orji, U., Ugwu, C., Asogwa, C.: An overview of quantum cryptography and shor's algorithm. Int. J. Adv. Trends Comput. Sci. Eng **9** (2020)
32. Grassl, M., Langenberg, B., Roetteler, M., Steinwandt, R.: Applying grover's algorithm to aes: quantum resource estimates, pp. 29–43. Springer, Berlin (2016)
33. Duan, B., Yuan, J., Yu, C.-H., Huang, J., Hsieh, C.-Y.: A survey on hhl algorithm: from theory to application in quantum machine learning. Phys. Lett. A **384**, 126595 (2020)
34. Hu, F., Wang, B.-N., Wang, N., Wang, C.: Quantum machine learning with d-wave quantum computer. Quantum Eng. **1**, e12 (2019)
35. Menneer, T., Narayanan, A.: Quantum-inspired neural networks **95**, 27–30 (1995)
36. Markov, I.L., Shi, Y.: Simulating quantum computation by contracting tensor networks. SIAM J. Comput. **38**, 963–981 (2008)
37. Blance, A., Spannowsky, M.: Quantum machine learning for particle physics using a variational quantum classifier. J. High Energy Phys. **2021**, 1–20 (2021)
38. Huang, Y., Jin, S., Zhang, Y., Pan, L., Shao, Q.: Quantum-inspired mean field probabilistic model for combinatorial optimization problems (2024). arXiv:2406.03502

Chapter 9
The Applications of Quantum Machine Learning in Today's World

Utkarsh Kedia, Sumedh Joshi, Trilok Nath Pandey, and Pankaj Shukla

Abstract This chapter provides an overview of the newly developed field of quantum machine learning (QML) and the diversification of its fields of uses. When integrated with the principles of quantum computing, and machine learning, QML provides enhanced computational power, which is likely to change the ways various sectors handle data and solve problems. The chapter also offers an extended discussion of the current applications of QML in finance, healthcare, cybersecurity and artificial intelligence. In applying QML to finance, it has potential in portfolio optimization, fraud detection and market prediction. In healthcare it holds promise in creating new drugs and therapies, in genetic diagnosis, and in medical imaging. The other areas of cybersecurity applications include improved methods of encryption, an anomaly detection system, and blockchain security. The chapter also highlights the application of QML in enhancing the natural language processing and reinforcement learning in artificial intelligence. Alongside with the general and specific benefits of the new field QML, the authors under consideration focus on the existing problems, such as the limitations of the hardware, algorithmic difficulties and integration problems. Finally, further considerations including future trends are reviewed based on the idea that enhancement of quantum hardware, algorithm, and interdisciplinarity would be crucial for the further development of QML in various areas.

Keywords Quantum machine learning · Quantum computing · Artificial intelligence · Finance · Healthcare · Cybersecurity · Portfolio optimization · Drug discovery · Encryption · Natural language processing

---

U. Kedia
School of Computer Science and Engineering, Vellore Institute of Technology, Chennai, India
e-mail: utkarsh.kedia2021@vitstudent.ac.in

S. Joshi · T. N. Pandey (✉) · P. Shukla
School of Electronics Engineering, Vellore Institute of Technology, Chennai, India
e-mail: triloknath.pandey@vit.ac.in

P. Shukla
e-mail: pankaj.shukla@vit.ac.in

U. Kedia · S. Joshi · T. N. Pandey · P. Shukla
School of Advanced Sciences, Vellore Institute of Technology, Chennai, India

S. Dehuri et al. (eds.), *Advances in Quantum Inspired Artificial Intelligence*, Intelligent Systems Reference Library 274, https://doi.org/10.1007/978-3-031-89905-8_9

9.1 Introduction

Quantum computing is a relatively new concept that uses skills of quantum mechanics in order to attain computations that are beyond standard computers. As opposed to classical bits, which are either 0 or 1's, quantum bits, or qubits, are in a state of superposition and hence, can simultaneously be 0 and 1 due to quantum effects such as superposition and entanglement. This makes quantum computers to able to work through various data sets in parallel and this can make them solve certain problems faster than classical computers.

Computer learning, also known as artificial intelligence, is a process with the help of which programs enable the computer to learn from the data provided and then decide or predict something on their own. Traditional or first generational Machine Learning (ML) algorithms comprising of decision trees, neural networks including artificial neural network (ANN), biological neural network (BNN) and convolutional neural network (CNN), support vector machines among others have been used with success in many fields such as image and speech recognition, natural language processing and predictive analytics to mention but a few. But the growing volume of data as well as its versatility can pose multiple problems to classical methods of ML, as the latter relies on the capacity of classical computers [1].

Bridging the Gap: A quantum machine learning method is another algorithm that has been introduced to be used with big data. quantum machine learning or (QML) combines the ability of quantum computing as a technique for machine learning. This is a branch of study that lies crossover, looking at the possibility of using the phenomena within quantum mechanics to make improvements to the performance of the ML algorithms. Relative to the classical ML, QML can provide ways of overcoming some of the major challenges that are associated with Big data in terms of efficiency in large and complex datasets [2]. Quantum machine learning, like other applications of quantum computers, has a number of advantages:

1. Exponential Speed-Up: Compared to classical computers, quantum computers can solve some problems in far less time allowing the training of the ML models to speed up as well.
2. Handling High-Dimensional Data: By its nature, QML algorithms can operate on high-dimensional data spaces and are thus recommended for higher order datasets that typically challenge classical algorithms.
3. Improved Optimization: Standard quantum algorithms are good looking for global optima in hardest probable optimization issues, which is beneficial for application such as portfolio optimization for stock brokers or for identification of probable drug content in medicines for the medical board [3] (Fig. 9.1).

9.1.1 Potential and Impact

The potential applications of QML span multiple industries, promising transformative impacts:

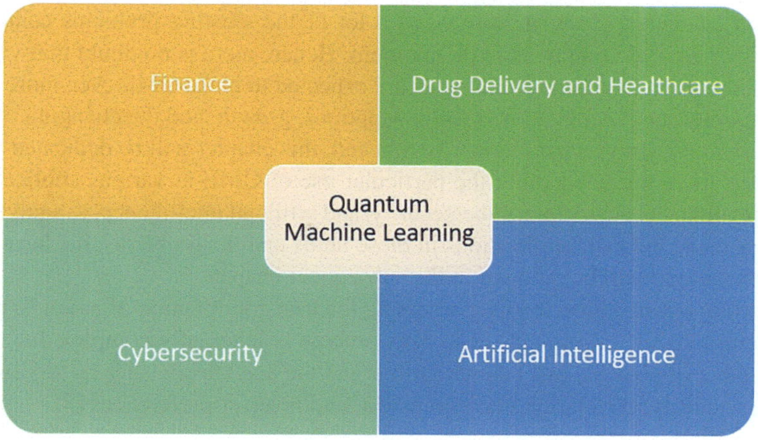

Fig. 9.1 Schematic representation of quantum machine learning bridging classical computing and quantum computing

- Finance: QML can superintend the improvement of portfolios, recognize dishonest operations, and presage tendencies more efficiently and effectively [1].
- Healthcare: The applications of QML include the possibility to transform drug discovery, develop better genomic analysis as well as improve the way medical imaging works in practice [4].
- Cybersecurity: These include improved encryption techniques, identification of network issues as well as the strengthening of security in block chain [5].
- Artificial Intelligence: The projected features can greatly enhance the developments of AI in the areas of natural language processing, reinforcement learning, image and speech recognition.

9.1.2 Present Problem/Questions and Future Dilemma/ Prospects

QML is a relatively new method, which explains why it does not fully meet all its proclaimed goals. Quantum computers presently are restricted in the number of qubits, error ratio and coherence time, preventing quantum computing from being used in practice. Although, current investigation and continuous changes in quantum hardware and even more in algorithmic progresses is gradually eradicating these problems. It is also anticipated that its deployment into real-world solutions will increase as the worlds of technology advances. There is research and development of the intermediate forms of quantum and classical computing or a combination of quantum and machine learning called QML for easier integration and more realistic workings in different industries [6] quantum machine learning is an emerging field that combines concepts of quantum computing and machine learning with the potential to change many industries, as a lot of the existing problems cannot be solved by classical computers at the moment. Hence, there is no doubt that various advancements in quantum technologies are expected to make QML even more valuable and relevant, paving the way for new options, growth, and the changing of data processing and analysis paradigms. To that end, this chapter will be dedicated to the review of literature concerning the particular use of QML in various applications, including finance, healthcare, cybersecurity, and artificial intelligence, to analyze the progress made in QML applications in these fields and the problems that have been encountered and will be outlined in the subsequent chapter.

Portfolio optimization involves selecting the best combination of assets to maximize returns while minimizing risk. This process is inherently complex due to the vast number of possible asset combinations and the need to consider factors such as market volatility, correlation between assets, and investor preferences [7].

9.2 Quantum Machine Learning in Finance

This section will outline the potential applications of Quantum ML in the financial sector and emphasise its impact in 3 major areas namely portfolio optimization, fraud detection and market prediction in detail [8].

9.2.1 Portfolio Optimization

Portfolio Optimization is a technique to select the optimal portfolio from a given set such that it maximizes returns while keeping the risk low. QML can be used to improve existing techniques as discussed below in detail [8].

9.2.1.1 Classical Challenges

While traditional optimization techniques such as MVO and CAPM work on the model, they have problems with computation when the portfolio size is large. These methods become inefficient with large numbers of assets, producing solutions that do not seek to extract maximum diversification advantage [9].

9.2.1.2 Quantum Advantage

Quantum algorithms like quantum annealing and quantum approximate optimization algorithm (QAOA) solve these problems by exploring a large number of asset portfolios in parallel. For example, Quantum Annealing finds application in solving combinatorial optimization problems best. With the help of superposition and tunnelling a quantum algorithm looks for the global optimum much better than a classical algorithm.

Example Quantum annealers sold by D-Wave Systems are reported in the literature to have been utilized in research to select optimal portfolios through the resolution of large-scale quadratic unconstrained binary optimization (QUBO) problems. This in turn can hold the ability to create better diversified portfolios for the investors with higher risk-adjusted returns.

9.2.2 Fraud Detection

Financial transaction security is important in preventing fraud and this makes fraud detection an important aspect for institutions and customers. This entails creating models that can point out abnormal patterns within the transactional data in relation to the swindle [10].

9.2.2.1 Classical Challenges

Traditional machine learning algorithms such as logistic regression and decision trees used for fraud detection cannot analyse huge amounts of transactions and need powerful computing resources. The training of these models can encounter a number of problems related to the accuracy of the results and the suitability of these models in the detection of new types of fraud.

9.2.2.2 Quantum Advantage

Quantum machine learning can be applied to upgrade present fraud detection platforms to become drastically proficient in pattern identification. QSVM and QNN can work well with high level data. Quantum algorithms are more effective at processing high-dimensional data. These algorithms can find visual correlations and recurring patterns in transaction data that could be unnoticed by classical ones.

Example A quantum-enhanced fraud detection system could predict and possibly control and estimate the level of suspicious activities within minutes from the transaction information on a daily basis with increased accuracy. This would also allow financial institutions to be better equipped to deal with threats as they occur and curb fraud cases.

9.2.3 *Market Predictions*

Market prediction is a process that entails the prediction of future prices and market trends with the help of the historical data and co-efficient derived from numerous economic indicators. Market forecasting, therefore, has several benefits, particularly to the traders, investors, and financial analysts. Table 9.1 provides a summary of the application of QML in finance [12].

9.2.3.1 Classical Challenges

For time-series forecasting, classical machine learning models are applied which includes autoregressive integrated moving average (ARIMA) and long short-term memory (LSTM) networks. However, such models are very computationally intensive and sometimes, do not fully capture the forces operating in the market.

9.2.3.2 Quantum Advantage

QML can bring a lot of value to the current market prediction models by implementing quantum computation strategies to work with the large sets of market data and find nonlinear relationships in them. Quantum-assisted time-series models can handle the data faster and at the same time produce better forecasts.

Example Scholars have attempted to apply quantum algorithms such as Grover and quantum Monte Carlo for finding a more efficient trading strategy that can forecast the stock prices and market conditions. The modern approaches being considered for enhancement of the accuracy of market forecasts are quantum Boltzmann machines and quantum-enriched reinforcement learning (Fig. 9.2).

Table 9.1 Classical and quantum-based methods in financial frameworks

Aspect	Classical approach	Quantum approach	Quantum algorithm
Portfolio optimization	Lower returns, Higher risk, Longer computation time	Higher returns, Lower risk, Faster computation time	Quantum Approximate Optimization Algorithm (QAOA)
Fraud detection	High accuracy, Slower computation	Higher accuracy, Faster computation	Quantum Support Vector Machine (QSVM)
Market predictions	Moderate accuracy, Higher error, Slower computation	Higher accuracy, Low error rates, and Low computation times	Quantum Neural Networks (QNN)

Fig. 9.2 Applications of quantum approach in financial applications

9.3 QML in Drug Discovery and Healthcare

Looking at opportunities, one can identify the application of quantum machine learning in Drug Discovery and Healthcare [12] (Fig. 9.3).

9.3.1 Drug Discovery

The process of drug discovery is long and multifaceted and includes recognizing possible drugs, creating substances, and evaluating their efficacy and toxicity. It can span years and billions of dollars and yet fail most of the time [1].

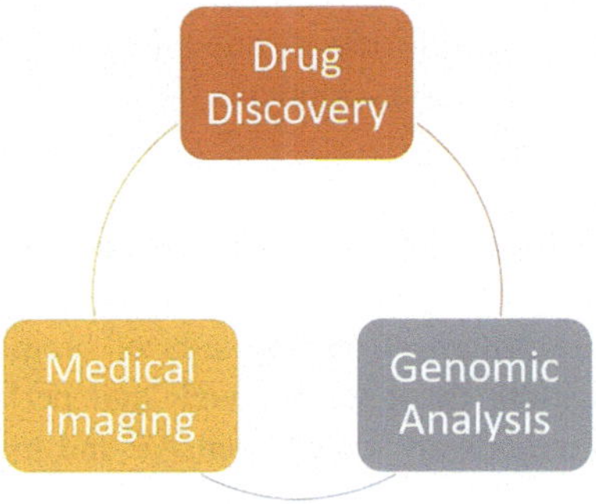

Fig. 9.3 Quantum machine learning in health care

9.3.1.1 Classical Challenges

Traditional and widely used methods for the analysis of the interaction between drug molecules and receptors include molecular docking and virtual screening, which are, as a rule, computationally intensive. These methods can be time-consuming and not very proficient when it comes to the search for new chemical compounds.

9.3.1.2 Quantum Advantage

Quantum computing could enhance healthcare by specifically simulating the process of molecular interactions that pertains to drug discovery; therefore, using quantum computing could be a more credible approach in finding suitable drugs. VQE and QAOA are among the quantum algorithms that can offer more accurate results in the optimization of molecular structures and prediction of binding affinities compared to classical approaches.

Example Scientists have proven that quantum computing holds the capacity to model challenging molecular structures like proteins ligand bond which is very essential in drug development. Such a capability could certainly cut down the time and the cost to introduce new drugs into the market.

9.3.2 Genomic Analysis

Genomic analysis is one of the research strategies that concerns the examination of an organism's genome, which includes all its genes and purposes to compare specific differences or variations of genes in individuals and their relationship to diseases. This field produces huge data which when analysed has to be done using complex computational methods [4–7].

9.3.2.1 Classical Challenges

Algorithms from the area of bioinformatics like sequence alignment and variant calling do not scale well and perform poorly on large genomic datasets. These methods are very expensive and may require large amount of time to analyse big data.

9.3.2.2 Quantum Advantage

There is the potential to improve genomic analysis by using QML for big genomic data processing based on quantum algorithms. Quantum computers have capabilities to work on multidimensional data and do data pattern matching and that's where they play the biggest role in decoding diseases associated with genes [9, 10].

Example Algorithms based on quantum computing can improve the pace of screening for the genes that are linked to certain diseases, thus improving diagnosis and possibly the beginning of a form of custom medicine.

9.3.3 Medical Imaging

MRI, CT, and PET are non-invasive techniques vital in the detection and treatment of several ailments. These technologies produce immense quantities of complex data which are not easy to analyse unless special methods are adopted for proper data analysis [13].

9.3.3.1 Classical Challenges

Techniques like CNNs that are used for image processing and analysis have seen a lot of success in the medical image analysis. However, these methods can be very time consuming and/or the optimization may not always converge to the 'best' solution, or the necessary accuracy cannot be reached in complicated and/or noisy images.

9.3.3.2 Quantum Advantage

QML can contribute an enormous amount to medical imaging since it advances the ways of perceiving images and analysing them as well. Quantum neural networks (QNNs) and other quantum-based image processing, in general, can efficiently pre-process medical images that will result in early detection and accurate diagnosis of various diseases.

Example Scholars are working on applying quantum-based image processing algorithms for increasing the accuracy of MRI & CT scan. This can help in the diagnosis of diseases such as cancer at its early stage and therefore treatment can be quickly administered (Table 9.2).

9.4 Quantum Machine Learning in Cybersecurity

Quantum machine learning in cybersecurity is an emerging synergy of quantum computing theories and sophisticated analytical methods. This growing field of study applies quantum algorithms for analysing and making sense of huge amounts of

Table 9.2 Comparison of classical and quantum techniques in healthcare applications

Aspect	Classical approach	Quantum approach	Quantum algorithm
Drug discovery	Time consuming and less efficient	Faster results with optimization	Variational Quantum Eigensolver (VQE), QAOA
Genomic analysis	Suboptimal performance on large datasets, Costly	Faster processing of big data leading to better analysis	Regular expression and other quantum algorithms
Medical imaging	Time consuming, Poor optimization	Enhanced image analysis	Quantum enhanced digital image processing

cybersecurity information, thus providing the basis of a new approach to threat identification and prevention. When implemented, quantum machine learning possesses incredible possibilities for enhancing computational capabilities which might alter the face of digital security as they are able to rely on two quantum mechanical phenomena; superposition and entanglement [5].

9.4.1 *Encryption and Decryption*

Encryption is critical in protection of data because it allows only the intended parties access to the information. Traditional encryption mechanisms depend on mathematical methods and computational problems that are taxing, for instance, the factorization of large numbers, or discrete logarithmic computations [5].

9.4.1.1 Classical Challenges

Existing cryptographic systems except for classical attacks essentially have large scale problems when exposed to a quantum attack. RSA and ECC and other conventional methods have been deemed safe from symmetric encryption algorithms and various other diverse methods. Nevertheless, these methods are insecure against quantum algorithms, especially against Shor's algorithm that can provide efficient solutions of mathematical problems, such as factorization and discrete logarithms. This quantum threat underlines the necessity of creating the quantum resistant cryptographic approaches for the long-term digital safety.

9.4.1.2 Quantum Advantage

Quantum machine learning (QML) encodes the capability to extend the creation of novel cryptography methods that are immune to both classical and quantum invasions. Quantum cryptography especially quantum key distribution (QKD) is a form of cryptography that uses principles of quantum mechanics to create fundamentally secure communications. Furthermore, the use of QML in cryptographic techniques can be improved because it is possible to exclude or optimize the key generation and other processes of encrypting information.

Example Scholars are working on creating quantum-safe techniques like lattice cryptography which are assumed to be resistant to the attacks from quantum systems. QML can improve these algorithms and make them fine-tuned with a potential of being implemented more frequently.

9.4.2 Anomaly Detection

This process is essential when it comes to determining the unfamiliar patterns of the network traffic, which, in turn, may manifest the presence of a threat, such as intrusions, viruses, or data thefts. Anomaly detection is an essential process that may include the assessment of big data to determine shift in behaviours from normal patterns in real time [5, 12].

9.4.2.1 Classical Challenges

It has been noticed that traditional machine learning algorithms for anomaly detection like clustering and neural networks are not designed to handle the volumes and intricate characteristics of network data. These methods can sometime take longer time to execute and can fail to capture very complex and minute anomalies.

9.4.2.2 Quantum Advantage

This in turn shows that QML can greatly improve anomaly detection when large scale network data is analysed using quantum algorithm. Since the quantum machine learning models are able to identify seemingly obscure patterns unlike the classical machine learning models, then this helps in early detection of the evasive cyber threats.

Example QSVM and QNN architectures can be applied to the network traffic data for analysis of anomalies with higher accuracy and increased speed. This in turn helps organizations to minimize on the time it takes to act on threats hence be in a position to respond to the threats appropriately.

9.4.3 Blockchain Security

In blockchain solutions applied to business, the transaction and data are protected by decentralized and unalterable databases. The security of the block chain networks is relatively depended with the cryptographic algorithm used to safeguard the efficiency of the transactions being processed in the network. Table 9.3 underlines the applications of the applications of QML in cybersecurity.

9.4.3.1 Classical Challenges

Despite the fact that at the classical level, the elements of the material are resistant to attacks, with the help of quantum computing, this situation changes significantly. Elliptic curve cryptography as well as other cryptographic algorithms that are being used by block chain may become vulnerable if controlled by a quantum computer.

9.4.3.2 Quantum Advantage

QML can improve the security of blockchain by designing new quantum-safe encryption methods and by improving blockchain algorithms. Quantum algorithms are capable of enhancing cipher processes for the stabilisation of blockchain systems in the quantum realm.

Example Currently, scientists are looking at ways on how quantum-resistant cryptographic algorithms can be implemented to blockchain platforms. Also, QML proves beneficial for enhancing the various consensus mechanisms and the processes of verifying transactions leading to the improvement of blockchain systems solidity and capability.

Table 9.3 Comparison of classical and quantum approaches in the context of cybersecurity applications

Aspect	Classical approach	Quantum approach	Quantum algorithm
Encryption and decryption	Susceptible to quantum attacks	Higher security using quantum principles	Quantum Key Distribution (QKD)
Cryptographic analysis	Limited by classical computational techniques	Enhanced cryptographic analysis capabilities	Shor's Algorithm, Grover's Algorithm
Anomaly detection	Slower and less accurate	Optimized and accurate detection with less false positives	Quantum Support Vector Machines (QSVM)

9.5 Advancements in Artificial Intelligence

This section highlights the impact areas of QML in the field of Artificial Intelligence (AI) such as Natural language processing and Reinforcement Learning. Table 9.4 provides a brief summary of the same.

9.5.1 Natural Language Processing (NLP)

Natural language processing is the ability of the computer and its programs to interface with the natural language of people. It makes it possible for the machine to comprehend, analyse, and then make co-ordinated actions towards what the human is saying or writing. NLP has use in uncovering translation services, in understanding polarity, and the creation of chatbots [4, 12, 13].

9.5.1.1 Classical Challenges

It is a fact that classical models of NLP, including recurrent neural networks (RNNs) and transformers, have advanced very much in the understanding and generation of human language. However, these models generally demand vast amounts of computational power and numerous datasets to obtain reasonably high accuracy levels. This is usually time-consuming, and can also be computationally demanding when fine-tuning the models.

9.5.1.2 Quantum Advantage

An example of how quantum machine learning (QML) can improve NLP is bringing in better training algorithms and inferencing. Quantum algorithms can keep up with the processing of large and complex data that are foundational in NLP, and hence, faster and more efficient models can be developed. The usage of quantum in NLP has

Table 9.4 Classical versus quantum approach in artificial intelligence application

Aspect	Classical approach	Quantum approach	Quantum algorithm
Natural Language Processing (NLP)	Resource intensive, Low efficiency	More efficient caused by faster training and inference	Quantum Boltzmann machines, Quantum transformers
Reinforcement learning	High time complexity, Computationally intensive	Improved learning and better optimization	Quantum annealing, Quantum enhanced policy optimization

the potential of enhancing tasks such as translation, summarization besides sentiment analysis.

Example The application of quantum Boltzmann machines and quantum-enhanced transformers can enhance the creation of the more efficient language models and accelerate the process of executing the NLP applications. This can lead to resolution of several issues including the development of better chatbots, accurate translation services and improved sentiment analysis [11–13].

9.5.2 Reinforcement Learning

Reinforcement Learning (RL) is one of the subcategories of Machine Learning techniques in which an independent agent forges a training technique that observes the environment in order to attain the maximum cumulative reward. Based on literature RL has seen substantial adoption in multiple real-life applications such as robotics, game playing and autonomous systems.

9.5.2.1 Classical Challenges

Most of the classical Reinforcement Learning algorithms including Q- Learning and Deep Q- Learning require balance between exploration and exploitation and is computationally intensive. For training RL models there is always timescale and computational issue involved particularly when the state and action spaces are high dimensional.

9.5.2.2 Quantum Advantage

Quantum RL can improve the optimization of learning and can also refine the volume and pace of the state-action searching processes through the help of quantum computation. There are also quantum algorithms like the quantum annealing and the quantum enhanced policy optimization which are capable of improving the convergence to the optimal policies.

Example Scientists are also trying to employ quantum RL in enhancing the working of automatic agents like self-driving car and robotic structures. The use of quantum enhancement in RL means that these systems can learn faster and make better choices allowing them to work as required in complex environments.

9.6 Challenges and Future Prospects

The field currently faces following major limitations namely hardware constraints, algorithmic limitations and limited integration with classical computing systems. These main concerns along with auxiliary ones are discussed below in further detail.

9.6.1 Technical Limitations

In this section we have discussed various technical limitations and challenges associated with this technology.

9.6.1.1 Quantum Hardware Constraints

Currently, one of the biggest issues relevant to QML is the existing state of quantum devices. While significant progress has been made, existing quantum computers are still limited by several factors:

- Qubit Quality: The main challenge of qubits is that they are difficult to stabilize because they are often coupled with decoherence and noise. A formidable technical challenge is the called decoherence and errors accumulation over longer periods of time.
- Scalability: Unfortunately, the great majority of quantum computers have a small number of qubits. Solving these issues and increasing the number of qubits from tens to hundreds or thousands, is a requirement in many real-world QML applications.
- Error Correction: It also highlights a rather serious issue that arises in quantum computation, namely the issue of quantum errors, for which reliable correcting methods, known as quantum error correcting codes, must be incorporated. The existing approaches to error correction are expensive and for generation of one logical qubit, a vast number of physical qubits are essential.

9.6.1.2 Algorithmic Challenges

Developing efficient quantum algorithms for machine learning tasks is another significant challenge:

- Algorithm Development: As a result, most of the classical machine learning algorithms do not have direct counterpart in quantum domain. Designing entirely new quantum algorithms which in turn out-perform their classical counterparts is a highly technical and iterative process.

- Optimization: Quantum algorithms frequently involve parameter optimization, the difficulties of which are born out of quantum measurements randomness and effective quantum to classical interface.

9.6.1.3 Resource Intensive

Both classical and quantum machine learning models can be resource-intensive, but quantum models face unique challenges:

- Quantum Simulators: The usage of quantum simulation of current algorithms on classical architectures is exponential with the number of qubits, thus rendering them infeasible for larger problems.
- Quantum Hardware Access: Quantum hardware is still scarce to this date and most of the time the costs are relatively high. Some quantum computation can be bought as a service from the Cloud; however, availability and cost are at present an impediment for most applications.

Data Requirements.

As it is with other machine learning models, whether they are highly advanced quantum ones or not, data, or rather training data, present a materials requisite for model training. Efficiently encoding and processing large datasets on quantum hardware is a significant challenge:

- Data Encoding: Classical data encoding into quantum states is not an easy process and could possibly be the weak link in QML methods.
- Data Processing: Quantum algorithms require a number of data read-outs essentially, which requires a substantial amount of computation followed by quartile coherence and error correction, thus it is challenging.

9.6.2 Integration with Classical Systems

In this section we have discussed how quantum technology can be integrated with classical system.

9.6.2.1 Hybrid Quantum–Classical Systems

Given the current limitations of quantum hardware, hybrid quantum–classical systems are a practical approach:

- Algorithm Integration: The nature of quantum and classical computations differs in a very significant way, which makes the development of the algorithms that are able to combine them rather complex. Such hybrid algorithms should incorporate

opportunities of both paradigm ideas although they possess different pros and cons.
- System Coordination: So, the exchange of data between the classical and quantum processors must be done effectively as any overhead will impact the overall performance.

9.6.2.2 Practical Implementation

Implementing QML in real-world applications involves several practical considerations: Implementing QML in real-world applications involves several practical considerations:

- Software Development: What is really needed is the development of software frameworks and libraries that help in embracing QML. The existing QPLs and QITs are still in their nascent stages.
- Interoperability: It is important to connect the classical and quantum systems, and also to have the ability to communicate the different quantum hardware platforms in the real-world applications (Fig. 9.4).

9.6.3 Future Prospects

This sub-section will discuss the future prospects and areas of work where changes will make a large impact on the field of QML as a whole.

9.6.3.1 Advancements in Quantum Hardware

Continuous advancements in quantum hardware will drive the future of QML:

- Improved Qubits: As the quantum computers' scale promotes, creating better quality of qubits with more coherence times and least error rates will be achievable

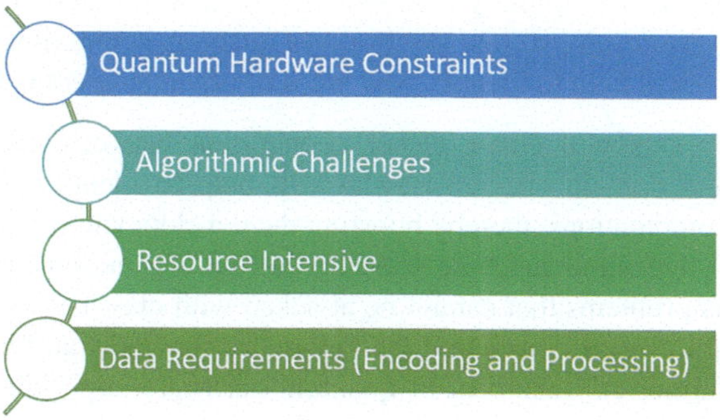

Fig. 9.4 Potential and weaknesses of quantum machine learning

- Scalability: Advancements of quantum chips and its deposition techniques to create larger quantum processors will be possible.
- Error Correction: As the technology of evaluating error rates develops, it becomes more possible to carry out calculations for longer and more complicated problems without significant errors.

9.6.3.2 Breakthroughs in Quantum Algorithms

Research in quantum algorithms will continue to expand the possibilities of QML:

- New Algorithms: Finding new quantum algorithms for machine learning that would be better solved using a quantum computer, as opposed to a classical computer, will be another big field.
- Optimization Techniques: The improvement of the quantum optimization methods will improve and broaden the routine applicability of QML models.

9.6.3.3 Integration and Standardization

Improved integration and standardization will facilitate the adoption of QML:

- Software Ecosystem: Developing the ecosystem of QML software tools and libraries will help make QML more available to the researchers and practitioners.
- Standards and Protocols: There is a necessity to set up the directions and guidelines for working with quantum computing and QML to easily integrate development workflows.

9.6.3.4 Practical Applications

As QML matures, its practical applications will expand across various fields:

- Industry Adoption: Finance, healthcare and cybersecurity industries, would continue to integrate QML solutions into their various operations because of the enhanced problem-solving capabilities.
- Interdisciplinary Research: To create more useful QML solutions, it is necessary to develop a close symbiosis of quantum computing specialists and end-users who will help to identify new areas of organizing computations.

This will create closer cooperation between quantum computing specialists and domain specialists who will help to identify new areas of organization of computations, thereby boosting the development of appealing useful approaches.

Quantum machine learning can be a game changer in multiple areas as long as problems that cannot be attacked with classical machine learning and have new abilities are considered. Therefore, we can find that, despite many existing problems, further enhancement of quantum hardware, algorithms, and interconnections will provide the foundation for QML's future practical applications and impact. Based

Table 9.5 Limitation, challenge and possibility of solutions on application of quantum machine learning in different fields

Topic	Limitations and challenges	Potential solutions
Finance	High noise levels in quantum systems affecting accuracy	Development of error-correcting codes and fault-tolerant quantum algorithms
	Scalability issues for large-scale financial data	Hybrid quantum–classical algorithms to manage data sizes
Healthcare	Limited number of qubits for complex calculations	Advances in qubit technology and quantum hardware scalability
	Quantum algorithms still in experimental stages	Increased research and collaboration between quantum computing and healthcare industries
Cybersecurity	Susceptibility of quantum cryptography to side-channel attacks	Implementation of robust quantum key distribution (QKD) protocols and continuous monitoring
	Quantum-resistant algorithms still under development	Focused research on developing and testing quantum-resistant cryptographic methods
Artificial Intelligence	High resource requirements for quantum AI models	Optimization of quantum algorithms and development of more efficient quantum hardware
	Difficulty in integrating quantum AI with existing systems	Creation of hybrid systems that leverage both quantum and classical computing strengths

on the evaluation of the nature of problems solvable with QML, there is strong outlook for future further development and exploring new opportunities in various areas (Table 9.5).

9.7 Conclusion

Quantum machine learning (QML) could be described as a new revolution in the ability of approaching and solving problems of different fields. By incorporating the quantum computing fundamentals with contemporary machine learning approaches, QML opens up the potential of highly improving the data analysis, increasing the efficacy of the complex systems, and identifying brand-new correlations in the enormous pools of the data.

Basically, QML has the capability to solve some of the most challenging problems in the area of finance including portfolio selection and management, fraudulent behaviour detection, and market forecasting that can outperform the classical models. It can be applied in the healthcare sector to quickly develop drugs, offer better genetic interpretation, and better picture interpretation, to increase patient quality of life and give precise treatment.

With the help of the proposed QML technique, cybersecurity presumably will receive better encryption, more efficient methods for identifying deviations from the norm, and advance blockchain defence against quantum and other types of invasions. In addition, the effect of QML on AI is great in as much as the enhancements of natural language processing which facilitate intelligent systems, reinforcement learning, and image and speech recognition for greater and more intelligent systems.

However, there is still quite a way to go in fully optimizing QML's potential, a fact complicated by certain difficulties. Loss and decoherence in quantum hardware, high computational demanding and challenges of interfacing with classical systems are well-known challenges in quantum computing. However, there is a great potential in QML's future in regards to the progressing quantum technologies, creation of unparalleled algorithms, and inter-disciplinary collaboration.

Future advancements in this domain will open new frontiers for the effective use of QML techniques. Quantum computing and machine learning can cooperate to open new horizons and create new opportunities and solve some of the challenging problems in the fields of science, Industry, and society. This new frontier is waiting for us to work on, and we are on the verge of a technological revolution that will shape the future of computing.

References

1. Canabarro, A., Mendonça, T.M., Nery, R., Moreno, G., Albino, A., Jesus, G.F., Chaves, R.B.: Quantum Finance: A Tutorial on Quantum Computing Applied to the Financial Market (2022)
2. Manjunath, T.D., Bhowmik, B.: Quantum machine learning and recent advancements. In: 2023 International Conference on Artificial Intelligence and Smart Communication (AISC), pp. 206–211. Greater Noida, India (2023). https://doi.org/10.1109/AISC56616.2023.10085586
3. Avramouli, M., Savvas, I., Vasilaki, A., Garani, G.: Unlocking the potential of quantum machine learning to advance drug discovery. Electronics **12**, 2402 (2023). https://doi.org/10.3390/electronics12112402
4. Davids, J., Lidströmer, N., Ashrafian, H.: Artificial intelligence in medicine using quantum computing in the future of healthcare. In: Lidströmer, N., Ashrafian, H. (eds) Artificial Intelligence in Medicine. Springer, Cham (2022). https://doi.org/10.1007/978-3-030-64573-1_338
5. Hdaib, M., Rajasegarar, S., Pan, L.: Quantum deep learning-based anomaly detection for enhanced network security. Quantum Mach. Intell. **6**, 26 (2024). https://doi.org/10.1007/s42484-024-00163-2
6. Flöther, F.F.: The state of quantum computing applications in health and medicine. Res. Dir. Quantum Technol. **1**, e10 (2023). https://doi.org/10.1017/qut.2023.4
7. Lin, W., Liu, H., Xu, J., Shi, L., Shan, Z., Zhao, B., Gao, Y.: Quantum machine learning in medical image analysis: a survey. Neurocomputing **525**, 42–53 (2023). https://doi.org/10.1016/j.neucom.2023.01.049
8. Maheshwari, D., Garcia-Zapirain, B., Sierra-Sosa, D.: Quantum machine learning applications in the biomedical domain: a systematic review. IEEE Access (2022). https://doi.org/10.1109/ACCESS.2022.3195044

9. Mironowicz, P., AkshataShenoy, H., Mandarino, A., Yilmaz, A.E., Luzern, H., Finanzdienstleistungen, I.F., Ifz, Z., Ankenbrand, S.T.: Applications of Quantum Machine Learning for Quantitative Finance (2024)
10. Rani, S., Kumar Pareek, P., Kaur, J., Chauhan, M., Bhambri, P.: quantum machine learning in Healthcare: developments and Challenges. In: 2023 IEEE International Conference on Integrated Circuits and Communication Systems (ICICACS), pp. 1–7. Raichur, India (2023). https://doi.org/10.1109/ICICACS57338.2023.10100075
11. Upama, P.B., Kolli, A., Kolli, H., Alam, S., Syam, M., Shahriar, H., Ahamed, S.I.: quantum machine learning in disease detection and prediction: a survey of applications and future possibilities. In: 2023 IEEE 47th Annual Computers, Software, and Applications Conference (COMPSAC), pp. 1545–1551. Torino, Italy (2023). https://doi.org/10.1109/COMPSAC57700.2023.00238
12. Cerezo, M., Verdon, G., Huang, H.Y., Coles, P.J.: Challenges and opportunities in quantum machine learning. Nat. Comput. Sci. **2**(9) (2022). https://doi.org/10.1038/s43588-022-00311-3
13. Chen, H.Y., Chang, Y.J., Liao, S.W. et al.: Deep Q-learning with hybrid quantum neural network on solving maze problems. Quantum Mach. Intell. **6**, 2 (2024). https://doi.org/10.1007/s42484-023-00137-w

Chapter 10
Unveiling the Role of Internet of Things (IoT) in the Landscape of Quantum Healthcare Monitoring

Sushree Bibhuprada B. Priyadarshini

Abstract In current era of technology, the adoration of Internet of Things (IoT) is rising day by day owing to extended internet connectivity as well as embedded technological proliferation. In this context, data processing and simulations could be greatly accelerated by quantum computing, which could have significant applications in the medical field. By analyzing large datasets, quantum computing may make it possible to create highly customized treatment regimens that take into account a patient's genetics, surroundings, and way of life. Fundamentally, IoT considers internet connectivity beyond the standard appliances like laptops, smart phones, tablets, desktops etc. with quantum monitoring for healthcare applications. Such devices can communicate as well as interact over internet and they can be tracked remotely. In this connection, IoT finds widespread applications in quantum biomedical engineering where the application of engineering principles as well as various design paradigms of biology are applied for various healthcare purposes. In this connection, quantum entanglement, in which particles become coupled in ways that affect one another instantly, may one day allow for instant communication or medical interventions, according to certain theoretical speculations. This idea is frequently investigated in relation to prospective medical advancements, despite the fact that it is highly hypothetical. This leads to closing the gap between engineering and quantum medicine. Moreover, miniaturization of several electronic devices together with the advancement of computer science and telecommunication give rise to novel quantum healthcare applications, thereby leading to outstanding revolution in the field of quantum medicine. Our chapter throws light on the latest illustrations of various implementations of IoT while addressing various technical issues associated with healthcare sector.

Keywords Data analytics · IoT · Quantum healthcare · Neuromechanics · Neuromodulation · Neuroscience

---

S. B. B. Priyadarshini (✉)
Institute of Technical Education and Research, Siksha 'O' Anusandhan University, Bhubaneswar, Odisha, India
e-mail: bimalabibhuprada@gmail.com; sushreepriyadarshini@soa.ac.in

S. Dehuri et al. (eds.), *Advances in Quantum Inspired Artificial Intelligence*, Intelligent Systems Reference Library 274, https://doi.org/10.1007/978-3-031-89905-8_10

10.1 Introduction

In the current era, internet is a network formed by inter-connecting computers throughout the world that provides a platform to communicate information over the network from one place to another. When we expand the capability of internet from connecting computers globally to the extent of connecting various small devices leading toformation of smarter technology called IoT [1, 2]. In this context, alternative medicine proponents have proposed ideas such as "quantum healing," which postulates that physical health might be impacted through quantum-level energy fields or awareness. Although these theories are debatable and unsupported by science, they are motivated by quantum mechanics and hold that both physical and energy states are related to health. For attaining this, the extremely sensitive gadgets known as quantum sensors are employed. These sensors might be able to identify minute alterations in biological systems, including variations in hormone or blood glucose levels that needs to be applied to non-invasive imaging procedures that provide more accuracy than conventional strategies. Moreover, it has been likely to be very soon the distinct things that we are seeing around us thatare all attending to be interconnected and has wide spectrum of applications in quantum healthcare and quantum physics.

Quantum physics is the foundation of quantum computing. Physical quantum phenomena like quantum entanglement and superposition are used in quantum computing. A quantum computer exploits a peculiar finding in quantum physics that denotes a single bit in both "1" and "0," referred to as a quantum bit or qubit. In essence, quantum computing uses this phenomenon to build a strong computing infrastructure that can process several pieces of data at once. This makes it possible to process enormous amounts of data in real time. As we move past the era of Moore's law, academics are increasingly interested in quantum computing as a way to advance computing capabilities in healthcare sector. In this context, the inability to determine the exact location of a spinning electron at any given time is known as quantum superposition. Conversely, it is computed as a probability distribution where the electron can exist with different probabilities at all times and places. Quantum computers work by using a collection of qubits in superposition to expedite operations and speed up computation. This technique is known as superposition in quantum health care.

Many compute-intensive healthcare applications are especially well adapted to quantum computing, particularly in the current highly connected digital healthcare paradigm, which includes networked medical appliances that can be connected to the cloud or the Internet. By 2025, it is anticipated that linked medical devices would generate 254.2 USD billion in sales, up from about 44.5 USD billion in 2018. Medical sensors, healthcare facilities, equipment, patients, physicians, and medical personnel are among the connected items. Monitoring and ensuring effective Quality of Services (QoS) across all linked infrastructures is one of the main problems in this heterogeneous connected paradigm.

Basically, what we enjoy services is a connection of distinct computers and computing appliances in case of quantum healthcare. Fundamentally, IoT interconnects different physical objects that we see around us like the air-conditioners, lighting-system in a room, fans [3] and anything and everything containing toothbrush, refrigerator, microwave ovens, so on. Not solely in our homes, however, additionally in our businesses like internetworking of various machines. The vision of internet of things is to interconnect these things that [1–5] we see in our homes, businesses places, offices, etc. The basic framework of IoT has been illustrated in Fig. 10.1 that basically operates through cloud computing paradigm and with assistance of various sensors that sense the ambience. The real striking question is how do we achieve interconnection of the things of daily use for quantum healthcare. These things are to be integrated with embedded electronics such that they have some fundamental computing platform attached to them and then they are going to be acting as distinct nodes of that interconnection.

Similarly, Fig. 10.2 illustrates IoT as an Interface between appliance and Human Value. Various processes like, device connection, data sensing, information exchange

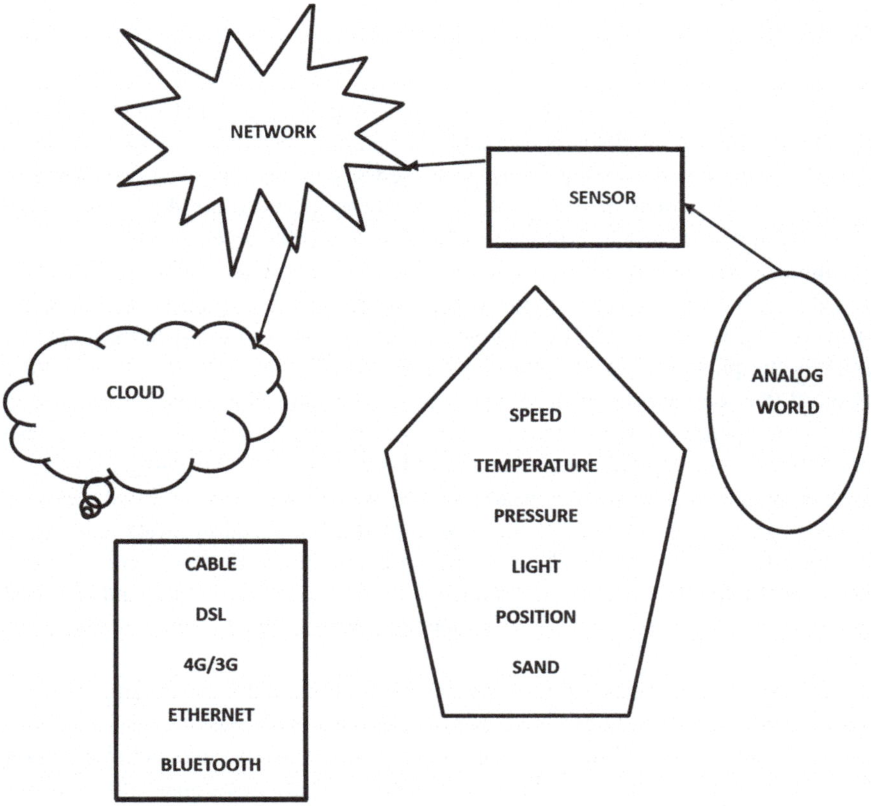

Fig. 10.1 Basics diagram of Internet of Things (IoT) and its operation

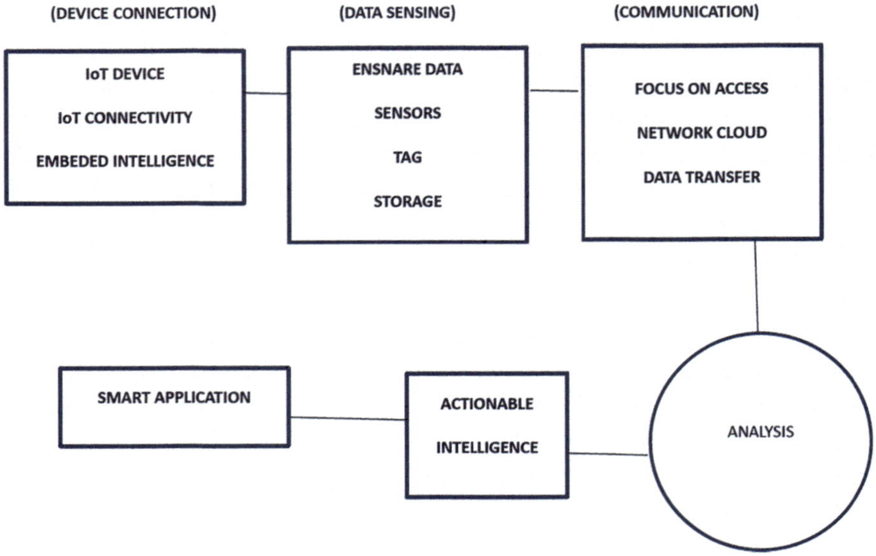

Fig. 10.2 IoT as an interface between device and human value

etc., take place in this context. Due to vast range of applications, IoT finds numerous applications in several fields particularly in quantum healthcare. Different kinds of applications emerge owing to usage of sensors in various ways as shown in Fig. 10.3.As illustrated in Fig. 10.3, for breathing purpose air flow sensors are employed. It operates on heat transmission as well as differential pressure.

Similarly, for patient position tracking, accelerometers are deployed, which is basically a dynamic sensor employed for vast range sensing purpose. Likewise, for measuring heart rate, electrocardiograms are employed. Electrocardiography is basically the phenomenon of recording the electrical activities of the heart over a period of time employing electrodes placed over the skin. Similarly, for measuring the body temperature, temperature sensors are deployed. Galvanic Skin Response Server (GSR) are applied for use in case of sweating [6]. The rising prevalence of persistent sicknesses such as heart sickness, kidney disorder, and diabetes give a large project to healthcare structures worldwide. Early and accurate analysis is vital for effective treatment and control of these conditions. However, conventional diagnostic strategies often rely upon a mixture of clinical know-how and standard laboratory checks, which may not be sufficient for well-timed detection, in particular in aid-constrained settings.

Machine learning gives a promising answer with the aid of leveraging massive datasets to identify patterns and make predictions that could assist medical selection-making. Despite its capability, several challenges prevent the substantial adoption of machine getting to know in disorder prognosis. These encompass the want for

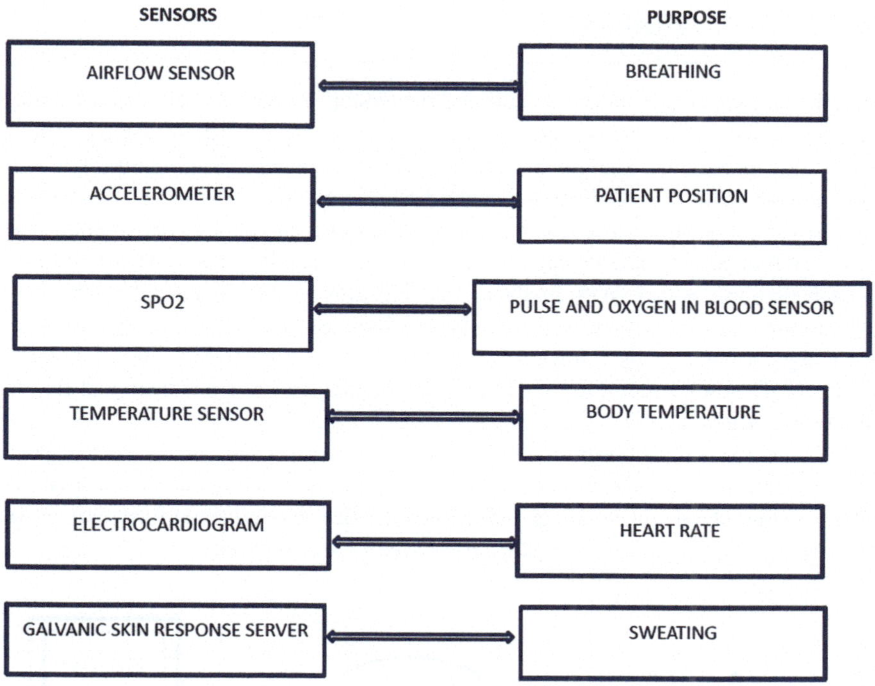

Fig. 10.3 Various sensors and their purposes in quantum healthcare

wonderful, various datasets, the interpretability of device gaining knowledge of fashions, and the mixing of these fashions into user-pleasant programs that can be without problems used by healthcare specialists and sufferers.

The rest portions of the chapter are arranged as follows: Sect. 10.2 details the diversified applications of IoT in quantum health care domain. The next section discusses the various elements associated with the IoT architecture and the biomedical implication aspects concerned with quantum healthcare. Section 10.4 describes the difference between IoT device administration and user management in health care framework. Subsequently, Sect. 10.5 shows the biological simulations with quantum healthcare. The enabling technologies for the quantum health care system are then discussed in Sect. 10.6. Afterwards, Sect. 10.7 brings the chapter to a close with a discussion on possible future works.

10.2 IoT and Its Applications in Quantum Healthcare

IoT can connect your fit bits to your vehicles, from your smartphones to the in-flight services, from home appliances like air conditioners to entire city.

10.2.1 IoT in Health Care

Most of us are already familiar with smart medical dispensers. This smart appliance helps patients by managing, storing, and dispensing their medications. This is a very tiny portion of a larger picture. Concerns about research, technology, and care may arise in the medical field and general practice. For medical examination, medical research must rely on leftover data in controlled environments. It is devoid of real-world data that can resolve pressing issues. IoT may have the solution to each of these issues. With analysis and real-time testing, it provides access to vast amounts of useful data. Additionally, IoT increases care quality and empowers medical personnel. Lastly, it lowers the medical devices' unaffordable high expenses. Figure 10.4 shows a scenario, where IoT and its medical uses has been justified in quantum healthcare.

IoT can also be helpful in remote health care tracking sitting at a distance place, which has been illustrated in Fig. 10.5, where a clinician (i.e. Doctor) can treat his patient sitting at a distance. He can also track the health condition of the patient. In this context, internet acts as the gateway between the two parties.

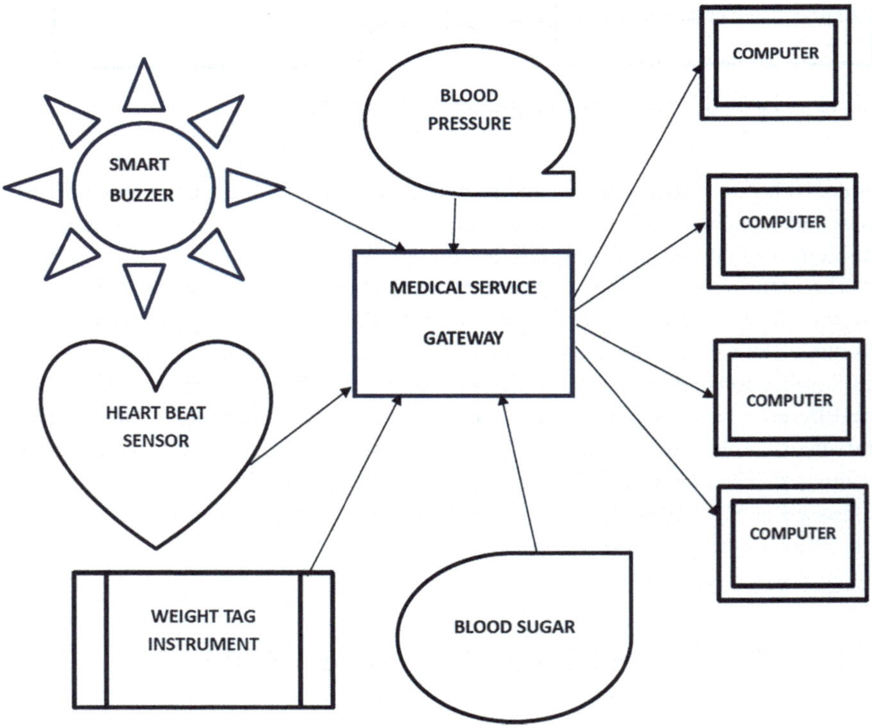

Fig. 10.4 IoT in affording medical services in quantum healthcare system

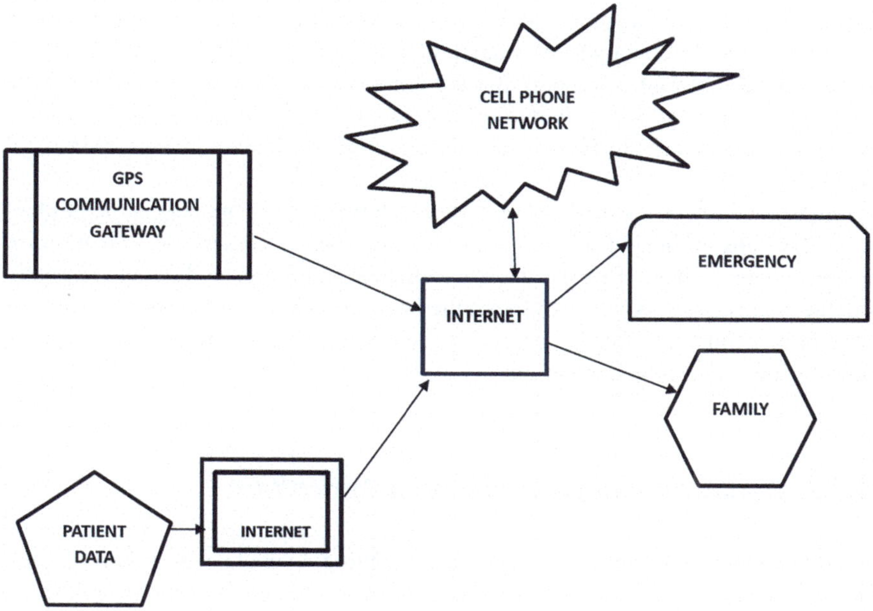

Fig. 10.5 IoT in remote health tracking in quantum healthcare system

10.2.2 Smart Thermostat

Let's see how internet of things will be helpful in alteration of temperature. A thermostat regulator is only a gadget to regulate the temperature. Your smart thermostat will be associated with the web. Also, it will be in sync with your smart phones. After installation of the thermostat, every time we alter the temperature, it will conserve the data. Often it will learn how you change the temperature. It will keep the distinct patterns of at the point when and how much temperature were desired. If there are sudden changes of weather, it will get the information from the internet and make the change in temperature in the blink of an eye. It senses the presence or absence of people in the home and automatically turn off air conditioners if no one is there in this way the electricity bill is reduced. It will find out whenever we route to the home, it will automatically turn on the AC's, so that you don't have to wait for cooling the surroundings. Smart! Isn't it? This is how IoT can be used to save energy and your precious money.

10.2.3 Waste Management

Imagine an area full of overfilled garbage containers that causes spillage all over place spreading diseases. Thanks to the Internet of Things. How? An IoT device is placed inside the container. A wireless sensor lets us know in real time the exact level of waste in the container. We may think it is expensive? Think again. We think the ecological footprint is too big? But the thing is sheer the reverse. The Proximal IoT platform gives us access to the LoRa network which consumes far diminished energy as compared to the 4G network. Furthermore, the battery of sensors lasts for 5–8 years [6, 7].

Imagine we can check the level of waste in all the city's containers in the blink of an eye. By transmitting out collecting vans to only where necessary, one can optimize the waste management. This will increase efficiency and reduce the fuel conjointly with the energy costs. Apart from these, the city's containers will always be emptied at the proper timeline. This is just one of the feasibilities on how IoT can improve current waste management solutions.

10.2.4 Smart Farming and Precision Agriculture

We all know how agriculture has become a gambling. Climate, water shortage, soil infertility stands as the crucial players in it. The use of internet of things will make agriculture beneficiary. Adequate water supply is essential for farming and lack of water will kill the Crops. Further, IoT improves water administration which becomes conceivable when combined with sensors, information and other apparatus. Such thing can be achieved by automated water sprinklers connect to IoT platform. Again, weather forecasting and other dynamic data inputs can affect crop productivity to a great extent. Further, IoT ensures accurate and effective communication to farmers to real time data related to agricultural process viz., weather forecasting, soil quality assessing, etc.

10.2.5 Smart Transportation

Vehicles are the biggest connected and computing devices with loT of sensors embedded to it. Vehicles generate huge data like travel time, origin, destination, vehicle volume, traffic movements, engine health, its surroundings, road conditions, etc. Further, such type of data is used by adaptive signal controls, engineering and construction projects that rely on traffic data collection. Again, vehicles communicate with each other and also with the surroundings, like traffic signals, roads, etc. to avoid traffic jams and damaged roads and also reach the destination swiftly. With smart parking, it is easier to find real time parking spot. IoT is an integral part of autonomous and self-driven cars.

10.2.6 Smart Retail

Now a days, Business are developing with innovative ways. Thus, functions the retail business organizations. Smart retail arrangements rearrange the way to better efficiencies, cost sparing, stock precision, more intelligent promoting and better client encounters. The utilization of RFID labels by retailers to build the productivity of their supply checks is the early case of IoT reception. Moreover, it enhances better stock administration by the utilization of sensors, guides, and RFID labels. The information is gathered on the cloud for examination and conveyed to the bosses. This empowers to screen what is accessible in the stock and in shops and take encouraging choices. It is extremely essential for enterprises like sustenance and pharmaceuticals which include steady observing of temperature, mugginess, and other ecological factors.

10.3 Components of IoT Architecture and Biomedical Implementation Aspects in Quantum Healthcare

When we come down to the IoT ecosystem, there's no single architectural design which is agreed universally [8, 9].

But there's a basic three level architecture that consists of:

(i) **Perception Layer**:
This is the bottommost layer of IoT ecosystem. In such layer, sensors actually gather the information from the environment around it. The sensors sense physical data or identifies objects in its surrounding. The data is then transmitted to the network layer.

(ii) **Network Layer**:
The network layer in itself takes up the charge of transferring this data from the sensors to the subsequent layer. This layer connects smart things, networking devices, servers and also processes sensor data [8–15].

(iii) **Application Layer**:
This is the topmost layer in IoT ecosystem. The main responsibility of the application layer is to deliver the information to the end user i.e. the patient of the health care system or the end platform. This layer can be aggrandized in to a 5-layer architecture as follows. It's quite similar to the above architecture [14, 15].

(iv) **Perception Layer**:
This layer remains the same where it collects information from the sensors from around its environment i. e. from the corresponding patient.

(v) **Transport Layer**:
This layer transport data between sensors to processing centre. This could be through wireless systems like Bluetooth, LAN, 3G, RFID, NFC or any medium we choose to.

(vi) **Processing Layer**:
Once the information has been transferred, the processing layer comes into picture where the huge amount of data is stored, analyzed and processed as per the user's requirement.

(vii) **Application Layer**:
This layer is same as the above 3 layered architecture. It is responsible to deliver various services to the end users.

(viii) **Business Layer**:
On top of all the layers stands the layer known as business layer. This layer can be used with any device when working on large scale environment. Moreover, business layer monitors the complete functioning of information collectors. For example, we can have this layers in various cars as well. If a vehicle is going to break down then we also attain the awareness with respect to the individual concerned vehicle and it also assists to reach out to the closest customer care center.

10.4 IoT Device Management Versus User Management and Health Care System

10.4.1 Device Management and Health Care

There will be around 50 million connected devices by 2020. As the IoT industry matures, over 88% of the companies see device management as an area of concern as it poses a threat to grow unless it is addressed.

i. **Device identification**: This process is performed to identify the device and to make sure that the device is genuine and it has reliable software and data.
ii. **Configuration**: This process adjusts the devices with the goal of the IoT system. During installation, some devices are written with fixed parameters like ID. Some have variable parameters that can be updated like time of data transmission of patient.
iii. **Diagnostics**: This procedure guarantees the efficient and safe performance of all devices in the network and decreases the risk of failures in case of health care systems.
iv. **Maintenance**: Such process incorporates novel functions, repairs security threats and fixes bugs in health care tracking.

a. **User management and Health Care**

Just like device management, it is necessary to have control over users accessing the IoT system. Further, user management recognizes users (i.e. patients), their duties, access type etc. Moreover, it's features allows to append and discard users, control over access levels and to perform operations in the system.

10.4.1.1 Bioinformatics

Data intensive, large-scale biological ultimatums of health care systems are considered from a computational view point. Basically, bioinformatics is a versatile field that discovers approaches and software tools for understanding biological information. As a many-sided field of science, bioinformatics incorporates information technology, mathematics, and engineering data while considering large-scale biological ultimatums from a computational view to analyze the biological data. Again, bioinformatics is viewed as both a crucial point for the collection of organic investigations which utilizing computer programming like a component of such approach, and in addition, a reference to specific investigation "pipelines" which get over and again is used, particularly in the field of genomics.

Bioinformatics and computational science incorporate the investigation of organic information, especially DNA, RNA, as well as protein groupings. The field of bioinformatics experienced unstable development beginning in the mid-1990s, driven to a great extent by the Human Genome Project and by quick peregrination in DNA sequencing innovation. Breaking down natural information to deliver significant data includes composing and running programs that utilizes calculations from diagram hypothesis, computerized reasoning, delicate figuring, information mining, picture handling, and PC reenactment. The calculations like this depend upon hypothetical establishments, for example, discrete arithmetic, control hypothesis, framework hypothesis, data hypothesis, and measurements.

10.4.1.2 DNA Sequencing

DNA Sequencing represents the process towards finalising the accurate demand of nucleotides along chromosomes and genomes. This assimilates the method or innovations which is used to decide the demand concerning four bases—adenine, guanine, cytosine, and thymine—within the strand of DNA. Besides, the emerging of fast DNA sequencing strategies has incredibly fasten organic and beneficial investigation as well as discovery. Study of DNA successions has emerged to be critical for essential research in health care systems and in several related fields of health research viz., restorative searching, biotechnology, virology, measurable science, etc. In case of chain-end DNA sequencing, the major DNA groupings have been obtained in mid 1970s by scholarly specialists employing difficult strategies dependent on two-dimensional chromatography. Following the advancement of fluorescence-based ordering techniques along DNA sequences, DNA sequencing has emerged as simpler and it demands of greatness as quicker as possible.

10.4.1.3 Genetic Engineering

Hereditary alteration/control as well as quality grafting are the terms which applies to the crucial regulation of a living being's qualities. In contrast to conventional rearing, a circuitous strategy for hereditary regulation along with hereditary designing employ present day instruments for sub-atomic cloning as well as alteration to particularly change the structure and attributes of target features. Moreover, hereditary building strategies have discovered accomplishment in distinct applications. Many authorities consider the enhancement of yield manufacturing (not a therapeutic implementation, however, relatively observe natural frameworks construction), the fabricate of engineered human insulin using adjusted microbes, the making of erythropoietin in hamster ovary cells, as well as the formation of novel sorts of trial mice, viz., the oncomouse (malignant growth mouse) for investigation.

10.4.1.4 Neural Engineering

Neural designing (also called neuro-engineering) representsa sequence inside biomedical building which employs designing methods to analyze, fix, supplant, update, or as a whole abuse the features of neural frameworks. Neural specialists are extraordinarily fit to deal with the structure issues at the interface of corresponding living neural tissue as well as for non-living objects. The principal diaries of investigation particularly gave to neural designing and universal gatherings on neural designing have been taken place by the IEEE, in Antalya, Turkey [11].

10.4.1.5 Neuroscience

Fundamentally, neurons represent the essential useful units of the sensory framework and are exceptionally specific cells which get equipped for sending such signs which work high as well as low level capacities required for existence as well as personal satisfaction.

Besides, neurons possess unique electro-preparation features which enable them to process data and subsequently, send that data to different cells. Moreover, neuronal action is needy on neural film potential as well as the progressions which prevails along and crosswise over this. Furthermore, a steady voltage, called the film potential, is typically kept up by specific convergences of particular particles crosswise over neuronal layers. Disturbances or varieties in this voltage make a lopsidedness, or polarization, over the film. Depolarization of such layer past its limit potential and produces an activity potential, which is the principal origin of flag transfer, called as neurotransmission of the sensory framework.

Again, an activity potential outcome within a course of particle transition over and over an atonal layer, making a powerful spike train or "electrical flag" that will transfer promote electrical alterations in different cells is considered. Stimulus can be manifested by electrical, compound, attractive, optical, as well as different types of improvements that impact the stream of charges, and consequently affects the levels of voltage crosswise over neural layers [11].

10.4.1.6 Neuromechanics

It represents the pairing of neurobiology, bio-mechanics, sensation as well as observation, and mechanical autonomy (Edwards 2010). Specialists are utilizing mobilized strategies and representations to examine the mechanical features of neural tissues and their impacts on the tissues' capacity to sustain and produce power conjointly with developments as well as in addition to the weakness to dread pack according to Laplaca and Prado. Such zone of investigation centers around to decode the changes of data among the neuromuscular as well as skeletal frameworks to create capacities and administering rules identifying with task and association of these frameworks as per Nishikawa et al. in 2007.

Furthermore, neuro-mechanics may be recreated by associating computational representations of neural circuits to representations of creature bodies that are arranged in logical and physical universes (according to Edwards). Moreover, test examination of biomechanics is carried out involving the kinematics and elements of developments. The procedure and examples of engine and tactile input amid development forms, and the circuit and physical association of the mind in charge of engine control are as largely as of now being explored for comprehending the volatility of creature development [12, 14, 15]. Moreover, researchers were engaged with the investigation of mechanical stretch of cell societies, shear disfigurement of planar cell societies, as well as shear misshapen of 3D cell comprising the networks. Comprehension of such procedures get trailed by advancement of working models fit for describing such frameworks under shut circle situations along extraordinarily characterized metrics.

The investigation of neuro-mechanics has gone for enhancing medicines for physiological medical issues which incorporates improvement of peg-leg structure, thereby carrying out the recycling of post development damage, and plan as well as regulation of versatile robots. Through considering the structures in 3D hydrogels, analysts will be able to distinguish new representations of nerve cell mechanical properties. Laplaca et al. in 2005 built up another model demonstrating that strain that may assume a job in cell culture [9–11].

10.4.1.7 Neuromodulation

Neuromodulation is aimed at treatment of disease along with injury by deploying medical equipment which can enrich or dominate the working of the whole nervous framework conjointly along the consignment of pharmaceutical entities, electric stimulus, or other articulation of energy signal to restore harmony in defective areas of human brain. Furthermore, investigators in such domain encounter the ultimatum of coupling suggestions in perceiving the neural stimulus to amelioration in technologies affording and assessing such stimulus with escalated susceptibility and vivacity in closed loop programs in the brain so that novel treatments conjointly with biological implementations can be manifested for treating such anguish from neural disturbances of different types [11, 12].

Moreover, neuromodulator appliances can rectify nervous system malfunction in connection to Parkinson's disease, tremor, chronic pain, OCD, Tourette's, severe abjection, epilepsy, etc. Neuromodulation is considered as a treatment for several deformities because it emphasizes on handling merely the highly specialized zones of the brain, conflicting the intrinsic treatments which may lead to many aftereffects on the body. Further, neuromodulator stimulators similar to microelectrode arrays are capable of stimulating and recording brain functionalities conjointly with further enrichments, that are aimed to become adaptable as well as reactive delivery appliances for drugs as well as other incentives [9–15].

10.4.1.8 Neural Regrowth and Repair

This uses neuroscience as well as engineering expertise for examining the surrounding as well as central nervous system work and for getting clinical outcomes to ultimatums manifested through brain functioning. Technologies employed to neuro-reproduction that emphasizes on technical appliances and substances; which afford the evolution of neurons aimed at particular implementations like the regeneration of peripheral nerve damage, the tissue regeneration for spinal cord injury, and for the reconstruction of retinal tissue.

Furthermore, genetic engineering as well as tissue culture represent the fields mellowing scaffolds for spinal cord to promote in assisting neurological disorders. Again, neuro-imaging strategies are employed for examining the hustle of neural frameworks, and the structure and activity of our brain. Neuro-imaging incorporates functional magnetic resonance imaging, Positron Emission Tomography (PET), Magnetic Resonance Imaging (MRI), as well as reckoned axial tomography (CAT) scans [11, 12, 14, 15].

Basically, functional neuro-imaging studies are focused where the zones of our brain carry out specialized works. Further, FMRI estimates hemodynamic task which is jointly associated with neural activity. This investigates the brain through converting the brain scanner to a fixed wavelength for perceiving such portion of our brain that gets actuated by carrying out distinct work through observing various things such as: PET, CT scanners, as well electroencephalography (EEG) that are recently being enriched and used for same type of works [10].

10.4.1.9 Brain-Computer Conjugation

These need to unswervingly interact with nervous system for tracking and prompting neural circuits while diagnosing and treating built in neurological non-functioning. Moreover, "Deep brain modelling" represents a crucial peregrination in such area which is efficient in treating mobility muddles.

10.4.1.10 Neuro-Robotics

Neurorobotics represents the review of the way the neural systems will be epitomized and mobilities imitated in automated devices. Neurorobots are specifically employed for knowing motor handling conjointly with mobileness, and memory determination, as well as value [10, 14, 15] frameworks and activity determination. By investigating neuro-robots in practical ambience those can be more lucidly marked and considering detail approaches of robot works indemnifying the concerned neural networks as well as sensitivity of such frameworks to its ambience [11, 12] are taken into account. The concerned framework models the brain attachment by employing a magnetic imaging resonance drawn out of a patient anguishing of so-called epilepsy [10, 14, 15]. Furthermore, such approach is capable of producing stimuli able to diminish the seizures.

10.5 Bridging Biological Simulations with Quantum Healthcare System

In current days, medical simulation has emerged to be a crying learning strategy by Lavoie and Clarke, 2017 and it details medical professionals those can simulate the scenarios which are not found during real-life patient care systems. Moreover, the learning strategy employed in the entire process of simulation has grown in due course of time, beginning out of a doll-like manikin along with simple functions to a practical manikin that get enriched through novel technologies incorporating embedded systems, robotics, and logical augmented reality, etc. Moreover, the usage of high definition simulation by Lavoie and Clarke in 2017 has been proven in the research to afford robust learning opportunities for the students while conjointly satisfying the particular learning needs.

Most of such frameworks are hybrid ones to map the real patient or medical ambience. In this context, many advanced technologies have been emerged via university research, specifically from technical schools, to promote the limits of such modeling requirements. Further, various companies manufacture products based on such technologies for manifesting an effectively smart environment for real life biological simulation. Although there is a necessity of conglomerating simulation technology with nursing curriculum, still there exist recently non documented technique of education for making the faculty prepared for efficiently operating and handling various simulation technologies.

Witness is there that assists the amelioration of simulation within nursing programs for escalating expertise of nursing students. The usage of Simulation Based Education has even proven as an evidence to escalate particularly the students' investigation scores. Huge number of researchers convey a lack of faith in connection to the domain knowledge of manikins employed in modelling applications. Further, an integrated study by Al-Ghareeb and Cooper proved huge number of proofs of obstacles to the conglomeration of modeling strategies as there owing to the complex technologies presiding in the engineering framework of the high-fidelity manikins, insufficient learning as well as inadequate provision to desired experts. The benchmark of use of modeling INACSL in 2016 [11, 12] in medical education is the precept where modelling is carried out by persons those have significant expertise and intelligence to carry out the educational opening worthy. A conglomerative redressal of healthcare study spanning from 2000 to 2015 relevant to HPSMs chooses the need for corresponding modelling coordinators and required level of technical assistance as a means of diminishing faculty embarrassment.

In this regards, on-site training has been the major constituent employed for training simulation technicians. Apart from this, few formal procedures for individuals exist to become efficient as a programming technician believed on educational strategy. Hurdles to modelling conglomeration persist to exist although the proof that SBE is emerging as a crucial discipline for applying in medical syllabus [10, 14, 15]. An obstacle identified the need for concerned persons those are able enough for affording desired technical expertise in healthcare sector. Figure 10.6 illustrates how to become successful by employing IoT solution in Quantum Healthcare System.

IoT has been conventionally more effective in technology peregrinations in contrast to the learning direction of medical strategy, as a training strategy for healthcare entities. Owing to fasten ameliorations in the IT sector, the so recent investigation has set objectives to merge such frameworks into the world of healthcare modelling. Numerous illustrations incorporate the formation of practical soft tissue enrichments for manikins which provide them more enrichment of frameworks which model haptic palpation, bile duct discovery, effective ultrasound modeling which can be incorporated with manikins. The progress of implementable framework for logical surgery modelling as well as heart modeling by Bramlet and others is also

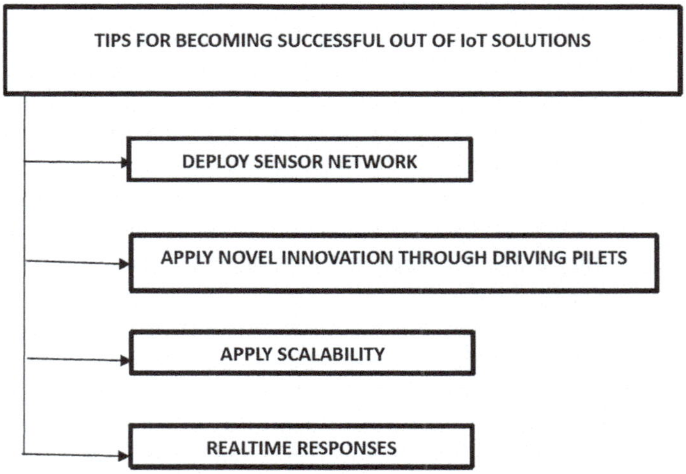

Fig. 10.6 Becoming successful from IoT solution

considered. The application of a comprehensive ambience which illustrates realistic surgical modelling and the amelioration of modern manikins which can model the human effect in response to the concerned curing process is also taken into account.

One of such modern robotic manikins employed in research is the SimMan3G proffered by Laerdal. Such manikin got constructed to be modelled to simulate effectively the regular human-illed ones with distinct health-concerned ultimatums and medical cases. Further, the hardware architecture of such robot is too much intellectual and comprises of five regulatable units. In such context, every module represents a different computer framework controlled by a principal computer, in which everyone uses a server framework for interconnecting to the outer environment. Hence, such manikin comprises of six computer modules and an inner router. Such modern framework is similar to the Electronic Control Units (ECUs) employed in the automobile area.

Moreover, the manikin possesses various relevant frameworks which can be regulated through such ECUs viz., the pupil dilation strategy, fluids pumping framework, breathing framework, conjointly with vibrators which map heart-beat conjointly with the seizure cases. Furthermore, this contains various sensors incorporating: radio Frequency determination sensors for medication usage, fluid rate tracking, and pupil dilation tracking. All such hardwares operate with modern software frameworks which are capable of modelling the ambience of a real patient in a real-life hospital using the patient control aspect, scenario programming tools, and various other software for affording functional regulations, multimedia data trapping as well as storing [15–22].

10.6 Quantum Healthcare and Enabling Technologies

This section introduces the quantum computing enabling technologies that facilitate the deployment of contemporary quantum computing systems. In particular, we outline our conversation by classifying the technologies that enable quantum computing into various domains, such as qubit technologies, hardware structure, control processor plane and host processor, quantum data plane, and quantum control and measurement plane [22–24].

10.6.1 Trapped Ion Qubits

In 1995, the first quantum logic gate was created using trapped atomic ions, which had been created using a theoretical framework that had been put out that year [22–25]. Qubit control has advanced technologically since its initial demonstration, opening the door for fully functional processors with a wide variety of quantum algorithms. Though trapped ions continue to be a significant obstacle, the small-scale experiment has produced encouraging results. The ions that function as qubits and a trap that integrates them in the appropriate places make up the trapped ion data plane. Different lasers are used in the control and measurement planes to carry out distinct tasks. For example, a precise laser source is utilized to alter the quantum state of a particular ion. A collection of photon detectors is used to measure the state of the ions by detecting scattered photons by them, and it also has a laser to cool and record measurements of the ions [24, 25].

10.6.2 Superconducting Qubits

Because of quantified states of electronic charge, these superconducting qubits exhibit quantitative energy levels when chilled to millikelvin temperatures. These are referred regarded be artificial atoms at times. They are an effective solution for quantum computing because of their compatibility with microwave control electronics, ability to use lithographic scaling, ability to operate at the nanosecond time scale, and constant improvement in coherence times.

10.6.3 Control Processor Plane

A suitable Hamiltonian or series of quantum gate operations and steps must be recognized and invoked by this plane for the control and measurement plane to carry out. These instructions execute the host processor's application to put a quantum algorithm into practice. The application ought to be specially constructed utilizing certain quantum layer functionalities provided by the software tool stack. The provision of a quantum error correcting algorithm is one of the control processor plane's most important duties. Various quantum processes needed for error correction are carried out using conventional data processing techniques based on calculated findings. The deduced delays could cause quantum computers to operate more slowly. If the time required for the error correction is comparable to that required for the quantum operations, the overhead can be decreased. In order to manage growing computational demands, this control processor plane would unavoidably have several interconnected processing units as the computational workload grows with machine size.

10.6.4 Quantum Control and Measurement Plane

The quantum plane's function is to transform digital signals that come from the control processor. It describes a collection of quantum operations carried out on the qubits within the quantum data plane. The data plane's analogue qubit output is effectively converted to classical binary data that the control processor can process with ease. Small qubit signals that cannot be corrected during an operation are caused by any variation in the signals' isolation, which leaves their corresponding qubit states with tiny inaccuracies. It is difficult to properly shield the control signals because they have to go via the device that separates the quantum data plane from the outside world.

10.6.5 Hardware Structure

Given that user data and network components associated with traditional computer systems are the focus of quantum computing applications. Consequently, it should be possible for a quantum computing system to effectively use conventional computer platforms. Qubit systems can be controlled using standard computer concepts, but they need meticulously coordinated control to function effectively. An analog gate-based quantum computer's hardware could be represented in several layers, such as the quantum data plane and control plane, to better understand the components that are required.

10.6.6 Quantum Data Plane

It is the primary element in the ecosystem for quantum computing. In general, it is made up of physical qubits and the structures needed to arrange them into a coherent system. It has the support circuits needed to determine the qubits' states and carry out gated operations. It accomplishes this for gate-based systems or for analog computers by managing the Hamiltonian [39]. A digital quantum computer's gate operations are controlled by control signals that are sent to the chosen qubits, which establish the Hamiltonian path. Sometimes two qubits are needed for gate-based systems, and the quantum data plane should have a programmable wiring network that facilitates communication between two or more qubits. The qubits provided by this layer must communicate more richly in analog systems. To achieve high qubit fidelity, strong isolation is necessary. Each qubit might not be able to communicate directly with every other qubit, which restricts connection. Consequently, computation must be mapped to certain architectural limitations that this layer provides. This demonstrates that the quantum data layer's connectivity and operation fidelity are its key features.

10.7 Conclusions

Though they sometimes create obstacles, the numerous cutting-edge technical elements give patients enormous opportunity to participate in real-life applications. Because many nursing faculty members are not very tech-savvy, when IT issues arise, they are unable to use the necessary and appropriate problem-solving techniques to fix and overcome technical hiccups. This means that a large number of the enormous simulation potential related to teachers and students are not utilized. To overcome obstacles to modeling-based education, a commendable approach is presented in this regard. Programs for healthcare simulation are also becoming more and more necessary in the market, as is the application of healthcare simulation technology. Maximizing the teaching tool's usefulness for medical practitioners has been impeded by a notably hurdle related to technological skills. Such barrier arises from a shortage of engineering experts with particular training in simulation to team with medical practitioners.

Acknowledgements The author is highly grateful to Department of Computer Science and Information Technology of Siksha 'O' Anusandhan University to provide required support for making this study successful.

References

1. Makhdoom, M.A., Abbas, H., et al.: Blockchain's adoption in IoT: the challenges, and a way forward. J. Netw. Comput. Appl. 251–279 (2019). Elsevier
2. Kang, B., Choo, H.: An experimental study of reliable IoT gateway. ICT Express **4**(3), 130–133 (2018). Elsevier
3. https://iot-analytics.com/10-internet-of-things-applications/. Accessed 30 Oct 2018
4. https://www.peerbits.com/blog/internet-of-things-healthcare-applications-benefits-and-challenges.html. Accessed 30 Oct 2018
5. https://www.recode.net/2015/1/15/11557782/a-beginners-guide-to-understanding-the-internet-of-things. Accessed 24 Nov 2018
6. https://imotions.com/blog/gsr/. Accessed 28 Nov 2018
7. https://www.edureka.co/blog/iot-applications/. Accessed 29 Nov 2018
8. https://aws.amazon.com/iot/
9. Sethi, P., Sarangi, S.R.: Internet of Things: architectures, protocols, and applications. J. Electri. Comput. Eng. **2017**, 1–25, Article ID 9324035 (2017)
10. Priyadarshini, S.B.B., et al.: The role of the Internet of Things (IoT) in Biomedical Engineering: Present Scenario and Challenges, pp. 1–274. Apple Academic Press, Taylor & Francis Group (2022)
11. Fitria, T.N., et al.: Internet of Things (IoT) in Education: Opportunities and Challenges. Prosiding Seminar Nasional Call Paper STIE AAS **6**(1) (2023)
12. Bornscheuer, U.T., Hohne, M.: Protein Engineering Methods and Protocols, Methods in Molecular Biology, vol. 1685, pp. 1–350. Humana Press, Springer Protocols (2018)
13. Omoniwa, B., Hussain, R., Javed, M.A., et al.: Fog/Edge Computing based IoT (FECIoT): architecture, applications and research issues. IEEE Internet Things J. **6**(3), 41184149 (2018)
14. Papp, C., Deeb, R.S., Booth, C., et al.: Bridging Medical Simulation with Computer Science and Engineering: A growing Field of Study, Nurse Education Today, vol. 71, pp. 1–6. Elsevier (2018)
15. Pervaiz, I., et al.: PRO-Net: a novel framework for augmenting android security against botnets and malware through advanced detection metrics. In: The Sciencetech, pp. 148–161 (2024)
16. Barry, D.W.: Role of health literacy screening in clinical practice—Response to Reddy et al's JAAD paper titled "Health literacy screening tools to identify patients at risk of misunderstanding wound care instructions after dermatologic surgery". J. Am. Acad. Dermatol. **90**(1) (2024)

17. Song, L., et al.: Smartphone-assisted ratiometric FRET aptasensor based on quantum dots and gold nano particles for point-of-care testing of zearalenone in cereals. Food Control **165** (2024)
18. Alrashed, S., Min-Allah, N.: Quantum computing research in medical sciences. Inf. Med. Unlocked **52** (2025) https://www.sciencedirect.com/science/article/pii/S2352914824001631
19. Rani, S., Kumar Pareek, P., Kaur, J., Chauhan, M., Bhambri, P.: Quantum machine learning in healthcare: developments and challenges. In: 2023 IEEE International Conference on Integrated Circuits and Communication Systems (ICICACS), pp. 1–7. Raichur, India (2023)
20. Hussain, M., et al.: A multi-objective quantum-inspired genetic algorithm for workflow healthcare application scheduling with hard and soft deadline constraints in hybrid clouds. Appl. Soft Comput. **128** (2022)
21. Qin, S., Pi, D., Shao, Z., Xu, Y., Chen, Y.: Reliability-aware multi-objective memetic algorithm for workflow scheduling problem in multi-cloud system. IEEE Trans. Parallel Distrib. Syst.Distrib. Syst. **34**(4), 1343–1361 (2023)
22. Chang, Y.W., et al.: Evolution of research subjects in library and information science based on keyword, bibliographical coupling, and co-citation analyses. Scientometrics, Springer Nat. **105**, 2071–2087 (2015)
23. Rehman, A., et al.: Emerging technologies for COVID (ET-CoV) detection and diagnosis: recent advancements, applications, challenges, and future perspectives. Biomed. Signal Process. Control **83** (2023)
24. Grumbling, E.: National Academies of Sciences, Engineering, and Medicine, Quantum Computing: Progress and Prospects. The National Academies, Washington, DC (2019). Available: https://www.nap.edu/catalog/25196/quantum-computing-progress-and-prospects
25. Rasool, R.U., et al.: Quantum Computing for Healthcare: A Review, pp. 1–24 (2021)

Chapter 11
Quantum Neural Networks: Exploring Quantum Enhancements in Deep Learning

Dukuru Chiranjevi, Anisha Mishra, Ranjan Kumar Behera, and Tarun Biswas

Abstract Deep Learning has demonstrated remarkable success in various domains, yet the computational demands of training large neural networks continue to pose challenges. This research investigates the integration of quantum computing principles into neural network architectures, aiming to explore and exploit the potential quantum advantages in deep learning tasks. The study focuses on Quantum Neural Networks (QNNs), where quantum bits (qubits) are leveraged to encode and process information in quantum superposition. Quantum entanglement and parallelism offer unique possibilities for enhancing the expressiveness and computational efficiency of deep learning models. We explore into the development of quantum-enhanced activation functions, weight encoding schemes, and novel layer structures adapted to quantum computing platforms. The research evaluates the performance of QNNs in comparison to classical deep neural networks across a spectrum of benchmark tasks, including image classification, natural language processing, and generative modelling. Metrics such as training convergence, model generalization, and computational speedup are analysed to measure the quantum advantage. Moreover, the investigation extends to hybrid quantum-classical approaches, exploring how classical and quantum components can simultaneously collaborate in the training and inference stages of deep learning tasks. This hybrid paradigm aims to control quantum speedup while maintaining compatibility with existing classical frameworks. This research bridges the gap between quantum computing and deep learning, providing valuable insights into the transformative potential of quantum-enhanced architectures for the next generation of AI systems. The outcomes of this study aim to guide

D. Chiranjevi (✉)
Department of Information Technology, Aditya University, Surampalem, Andhra pradesh, India
e-mail: chirudukuru2266@gmail.com

A. Mishra
Department of Computer Science and Engineering, KIIT University, Bhubaneswar, Odisha, India

R. K. Behera · T. Biswas
Department of Computer Science and Engineering, IIIT Ranchi, Ranchi, Jharkhand, India
e-mail: ranjan@iiitranchi.ac.in

S. Dehuri et al. (eds.), *Advances in Quantum Inspired Artificial Intelligence*, Intelligent Systems Reference Library 274, https://doi.org/10.1007/978-3-031-89905-8_11

future advancements in leveraging quantum computing for enhancing the efficiency and capabilities of deep neural networks.

Keywords Deep learning · Quantum computing · Neural networks · Machine learning

11.1 Introduction

The use of quantum circuits significantly enhances feature extraction, leading to improved classification performance [1]. However in the incorporation of quantum computing into machine learning comes with challenges, including noisy quantum hardware, qubit limitations, and complexities in training and optimization. These challenges are discussed within the broader context of quantum machine learning (QML) and quantum deep learning (QDL), highlighting how these fields leverage quantum computing's unique properties to address high-dimensional data analysis problems [2].

A primary goal of recent research is to design and analyze the Res-HQCNN model, which is capable of learning unknown unitary transformations from quantum data in both noisy and noise-free environments. Researchers consider the Res-HQCNN model a crucial advancement in the incorporation of quantum computing with neural networks [3]. Another study focuses on optimizing routing in quantum networks through deep reinforcement learning, leading to the development of the DQRA model. This model enhances secure and efficient data transfer by leveraging entanglement-based routing techniques, ensuring high security and reliability for future quantum networks [4].

The probable of hybrid quantum-classical neural networks in machine learning is also explored, particularly in image classification tasks. By comparing different quantum circuits and model configurations, researchers assess their efficiency in handling multiclass problems. Findings suggest that certain quantum circuits significantly boost classification performance [5]. In reinforcement learning, quantum algorithms provide a promising approach for enhancing decision-making in uncertain environments. A proposed quantum deep reinforcement learning framework based on variational quantum circuits demonstrates the advantages of quantum computing in high speed and complex decision making tasks [6]. Furthermore, the usage of graph neural networks (GNNs) in deep reinforcement learning for resource allocation in CFNs (Cloud and Fog Networks) has been investigated, aiming to optimize bandwidth and service scheduling [7].

Cybersecurity is another area where quantum neural networks (QNNs) show great promise. Given the increasing threats from malware attacks on smart grid components, researchers propose a quantum convolutional neural network (QCNN)-based approach for malware detection. This method enhances the detection rate by utilizing quantum feature extraction techniques, providing a more robust security mechanism compared to classical methods [8]. Furthermore, quantum machine learning addresses challenges of huge datasets in complex computations through methods such as quantum clustering and quantum-enhanced pattern recognition showing potential in artificial intelligence applications [9].

Several studies emphasize the incorporation of quantum computing into reinforcement learning, addressing inefficiencies in existing algorithms. Despite current limitations, such as vanishing gradients and limited qubit availability, new distributed algorithms are being examined to harness the strength of near term quantum computing [10]. The study of quantum enhanced anomaly detection frameworks demonstrates how quantum autoencoders can improve network security by accurately identifying cyber threats [11]. Additionally, the scalability of deep neural networks (DNNs) for big data applications is analyzed, with discussions on leveraging GPUs and neuromorphic computing to enhance processing efficiency [12].

The major contributions in this chapter are as follows:

- We have presented an in-depth discussion of quantum machine learning, quantum deep learning, and quantum neural networks from 2017 onwards, covering fundamental principles, applications, characteristics, models, and challenges.
- Some of the key challenges in the field, such as scalability, noise in quantum hardware, and optimization difficulties are discussed, while identifying general strength opportunities for future advancements.
- The study explores promising research directions in quantum computing, including advancements in quantum algorithms, hybrid quantum-classical models, and scalable quantum neural network architectures.
- Some practical usage of quantum neural networks over several domains, such as cybersecurity, healthcare, and high-dimensional data analysis, are examined, emphasizing their potential transformative impact.

The rest of the paper is organized as follows: A Systematic literature review of the related work are presented in Sect. 11.2. The Fundamental of quanatum machine learning along with various applications are discussed in Sect. 11.3. The main role of neural network in quantum computation along with embedding of quantum state in neural network is presented in Sect. 11.4. several challenges and opportunities in the area of quantum neural network is presented in Sect. 11.5. the further possible research directions are also discussed in Sect. 11.6. Finally, the work is concluded in Sect. 11.7.

11.2 Related Work

The incorporation of quantum computing with machine learning, particularly in domain of image classification and deep learning, has garnered significant attention in recent years. Golchha et al. [13] proposed a new technique to multiclass image classification by incorporating Quantum Convolutional neural networks (QCNN) into Convolutional neural networks (CNN). This approach leverages the principles of quantum mechanics, such as superposition and entanglement, to improve the performance of traditional CNN architectures in image processing tasks. The study demonstrated that quantum computing methodologies could significantly increase the accuracy and efficiency of image classification tasks, highlighting potential of quantum-enhanced machine learning models.

Yosif et al. [1] further explored the capabilities of QCNNs, emphasizing their potential to reduce computational load and processing time by optimizing specific parameters. However, the study also identified several challenges associated with integrating quantum computing into deep learning, such as high error rates in qubits and the complexity of designing quantum circuits. These challenges are particularly pronounced in medical image classification, where classical computational methods often struggle with large datasets and complex computations. Despite these hurdles, the authors underscored the transformation power of quantum computing in overcoming the limitations of classical approaches. Priyanka et al. [2] provided a comprehensive analysis of quantum machine learning (QML) and quantum deep learning (QDL), highlighting the advantages of quantum models over classical ones. The study discussed fundamental principles of quantum computing, how QML can outperform traditional machine learning in certain scenarios. However, the authors also acknowledged the practical obstacles to implementing these technologies, such as the necessity in robust quantum hardware and error correction mechanisms.

Liang et al. [3] introduced Res-HQCNN, a hybrid quantum classical neural network model that leverages deep residual learning to enhance the learning capabilities of quantum neural networks (QNNs). This model is particularly effective for handling both clean and noisy quantum data, providing new directions for further research in quantum machine learning. The study emphasized the importance of developing practical quantum algorithms that can be used to real-world information, paving the way for various robust and scalable quantum machine learning models. Le et al. [4] proposed a novel approach to entanglement routing in quantum networks using machine learning techniques. The authors introduced the deep quantum routing agent (DQRA), a deep reinforcement learning model that optimizes routing paths by considering quantum entanglement. DQRA employs a deep neural network to order traffic requests and a qubit-preserved shortest path algorithm for routing, demonstrating the potential of machine learning to enhance quantum network efficiency. Incudini et al. [14] discussed the quantum path kernel (QPK), a generalized neural tangent kernel that characterizes the dynamics of quantum neural networks (QNNs) in the factors of deep quantum machine learning. The study highlighted the need for frameworks that address the quantum advantages and drawbacks of existing models, such as the barren plateau problem in QNNs. This work underscores the importance of developing theoretical foundations for quantum machine learning to ensure its practical applicability. Trochun et al. [5] investigated the interpretation of hybrid quantum traditional neural networks in multiclass classification , including image classification using the MNIST and Fashion MNIST datasets. The study demonstrated that quantum circuits could improve the performance of classical neural networks, suggesting that future work should focus on optimizing quantum circuits for various learning tasks.

Lokes et al. [6] explored the application of quantum deep reinforcement Learning (QDRL) with variational quantum circuits (VQCs) to solve various types of challenges that are challenging for traditional computers. The study highlighted the potential of quantum computing to enhance reinforcement learning algorithms, Specifically in the factors of Noisy Intermediate Scale Quantum (NISQ) devices. This work opens new avenues for research in quantum-enhanced reinforcement learning. Han et al. [7] proposed a novel approach to resource allocation in computing force networks (CFN) by combining graph neural networks (GNNs) with deep reinforcement learning (DRL). The study emphasized the importance of real-time application development and the need for efficient service scheduling and bandwidth management in cloud and edge computing environments. This work highlights the power of quantum enhanced machine learning to manage complex resource allocation challenges. Akash et al. [8] introduced a cloud based malware identification framework for smart grid devices using Quantum Convolutional Neural Networks (QCNNs) and Deep Transfer Learning (DTL). The study addressed the growing cybersecurity challenges associated with smart grid technologies, demonstrating the potential of quantum-enhanced machine learning to improve malware detection in critical infrastructure. Chen et al. [15] proposed the QDQN-DPER method, which combines quantum deep Q-learning with distributed prioritized experience replay to optimize the learning of quantum reinforcement learning (QRL) agents. The study demonstrated that the proposed approach outperforms traditional quantum deep Q-learning techniques, achieving higher and more stable scores across various tasks.

Gupta et al. [9] explored the intersection of quantum computation and machine learning, highlighting how quantum algorithms can enhance artificial intelligence (AI) and deep neural networks. The study discussed the potential of Quantum Neural Networks (QNNs) to process complex data more accurately than traditional methods, leveraging quantum properties such as superposition and entanglement. Kwak et al. [16] provided an overview of quantum deep learning, emphasizing the usage of quantum computing methods to improve the training of deep neural networks. The study highlighted the several applications of Quantum Neural Networks (QNNs) and the challenges associated with this emerging research field. Manjunath et al. [17] proposed an innovative approach to improving Deep Q-Learning (DQL) using Parameterized Quantum Circuits (PQCs). The study demonstrated that quantum enhancements in DQL could lead to faster convergence and higher rewards estimated to traditional DQL approaches. Padha et al. [18] discussed the integration of quantum computing principles into machine learning, particularly in time series analysis. The study demonstrated that quantum-enhanced models, such as Quantum Convolutional Neural Networks (QCNNs), Quantum Recurrent Neural Networks (QRNNs), and Quantum Long Short-Term Memory (QLSTM) models, outperform classical models in time series analysis tasks. Bova et al. [19] examined the several applications of quantum computing in various sectors, highlighting its ability to solve difficult computational and combinatorial problems. The study emphasized the importance of advancing quantum technology to address practical challenges in commercial scenarios.

Krelina et al. [20] explored the usage of quantum technologies in military applications, discussing the main benefits and challenges related with their implementation. The study highlighted the need for a skilled workforce and a supportive quantum ecosystem to realize the full strength of quantum technologies in defense. Moussa et al. [21] addressed the challenges of parameter tuning in the Quantum Approximate Optimization Algorithm (QAOA) using unsupervised machine learning techniques. This study demonstrated that these techniques could effectively optimize parameters for QAOA, avoiding the difficulties associated with traditional optimization methods. Li et al. [22] proposed a quantum algorithm for data classification based on the nearest neighbor algorithm. This work demonstrated that quantum circuits could be used to classify data more efficiently by dividing data into smaller subsets and creating integrated quantum circuits for classification. Higham et al. [23] proposed a method for transferring trained artificial neural networks into a quantum computing environment using quantum annealing. The study demonstrated that this approach could significantly improve classification times, highlighting the potential of quantum optimization techniques. Yu et al. [24] proposed a quantum algorithm for secure deep neural network (DNN) predictions, leveraging quantum cryptographic capabilities to enhance data confidentiality and integrity. The study highlighted the power of quantum computing to address the limitations of traditional data protection methods. Wen et al. [25] proposed an algorithm that improve the expressivity of quantum neural networks by adding auxiliary qubits to the data encoding and trainable blocks of QResNets. The study demonstrated that this approach could improve the production of quantum neural networks in various tasks.

Mellak et al. [26] proposed a deep neural network model for generating non-equilibrium steady state solutions for correlated open quantum many body systems. The study demonstrated that this approach could enhance previous methods, such as neural density operators (NDO) and restricted Boltzmann machines (RBM). Chen et al. [27] proposed a hybrid quantum neural network for deep reinforcement learning, demonstrating that variational quantum circuits (VQCs) could improve the training of agents in maze-solving tasks. The study highlighted the capacity of quantum computing to enhance traditional learning algorithms. Kawase et al. [10] proposed a distributed architecture for quantum neural networks (QNNs) to address problems such as vanishing gradients and limited expressibility. The study demonstrated that this approach could improve the performance of QNNs on current quantum devices. Hdaib et al. [11] explored the use of quantum machine learning (QML) and quantum deep learning (QDL) for detecting anomalies in network traffic. The study proposed three novel approaches that incorporate quantum autoencoders with Q-OC-SVM, Q-RF, and Q-KNN classifiers, displaying the strengths of quantum computing to enhance cybersecurity. Ratnaparkhi et al. [12] provided an audit of the new developments in deep neural networks (DNNs), emphasizing their applications in speech recognition, image processing, and language understanding. The study highlighted the need for scalable platforms to support the training and deployment of DNNs in various domains.

A detailed and comprehensive analysis of various research works related to Quantum Neural Networks is presented in Table 11.1. The literature survey highlights the growing interest in incorporation of quantum computing with machine learning, particularly in image classification, deep learning, and reinforcement learning. Quantum-enhanced models, such as Quantum convolutional neural networks (QCNNs) and Quantum neural networks (QNNs), have demonstrated significant potential in improving the achievement of classical machine learning algorithms. However, several challenges remain, including the high error rates in qubits, the complexity of designing quantum circuits, and the need for robust quantum hardware.

Future research directions should spotlight on expressing these tests by developing more efficient quantum algorithms, optimizing quantum circuits, and probing new appliance of quantum machine learning in various sectors. Additionally, there is a need for theoretical frameworks that can guide the real time implementation of quantum improved machine learning models. As quantum computing grows to evolve, it is expected that these advancements will lead to more robust, scalable, and accurate machine learning solutions, making the way for life changing applications in artificial intelligence and beyond.

Table 11.1 Some of the potential existing Research works on Quantum Neural Networks

Authors	Year	Key contributions	Domain	Challenges addressed
Golchha et al. [13]	2023	Demonstrated higher recognition accuracy and lower training loss in QCNN-enhanced CNNs	Image processing	Limitations of classical CNNs in achieving efficiency during classification
Yousif et al. [1]	2024	Developed a quantum circuit for convolution and pooling layers achieving better accuracy and speed	Medical image analysis	High error rates in qubits and designing quantum circuits
Priyanka et al. [2]	2023	Synthesized findings on QML and QDL advantages like speed optimization and quantum parallelism	Quantum machine learning	Complexity and scalability of conventional ML algorithms

(continued)

Table 11.1 (continued)

Authors	Year	Key contributions	Domain	Challenges addressed
Liang et al. [3]	2021	Introduced Res-HQCNN for learning unknown unitary transformations in noisy and clean quantum data	Quantum data processing	Inefficiency of conventional quantum data transformation methods
Le et al. [4]	2022	Proposed DQRA leveraging deep reinforcement learning for optimized routing in quantum networks	Quantum networking	Scalability and routing challenges in qubit-scarce environments
Incudini et al. [14]	2023	Bridged deep learning and quantum computing through hierarchical parametric dependencies in QNNs	Deep quantum learning	Barren plateau problem and convergence challenges in QNNs
Trochun et al. [5]	2021	Multiclass classification evaluation assessment of quantum circuit topologies	Quantum machine learning	Exploration of quantum circuit efficiency and difficulties with multiclass classification
Lokes et al. [6]	2022	Development of VQ-DQN resource efficiency performance in diverse environments	Quantum machine learning and reinforcement learning	Memory consumption and resource utilization implementation on NISQ devices
Akash et al. [8]	2023	Novel framework design utilization of quantum convolutional neural networks	Cybersecurity in smart grid technologies	Increased malware vulnerability malware detection need for device-specific solutions
Chen et al. [15]	2023	Assessment of contemporary quantum algorithms performance improvement	Quantum reinforcement learning	High resource demand for training inefficiencies in existing training techniques

(continued)

Table 11.1 (continued)

Authors	Year	Key contributions	Domain	Challenges addressed
Gupta et al. [9]	2017	Quantum neural networks and clustering	Quantum machine learning	Time complexity issues potential for AI advancement
Kwak et al. [16]	2021	Quantum Neural Networks (QNNs)	Quantum deep learning	High constraints of classical deep learning techniques future challenges in advanced quantum algorithms
Manjunath et al. [17]	2024	Quantum enhancement of DQL improvements in performance metrics	Quantum machine learning	Quantum circuit design noise and error mitigation integration with classical frameworks
Padha et al. [18]	2024	Development of quantum enhanced models	Quantum machine learning	Complexity of quantum models ntegration of quantum and classical approaches
Bova et al. [19]	2021	Industry application insights enhanced prediction and decision making	Quantum computing	Optimizing decision-making processes complexity of data assessment scalability of solutions
Krelina et al. [20]	2021	Cautions against quantum hype opportunity and threat analysis implementation challenges	Military applications of quantum technologies	Establishing a national quantum environment overcoming implementation and transition from experimentation to application delays
Moussa et al. [21]	2022	Parameter optimization unsupervised machine learning techniques	Quantum Approximate Optimization Algorithm (QAOA)	Tuning parameter difficulties complexities involved in implementing unsupervised machine learning

(continued)

Table 11.1 (continued)

Authors	Year	Key contributions	Domain	Challenges addressed
Li et al. [22]	2021	Two-step classification process computational efficiency quantum circuit integration	Quantum computing and machine learning	Complex quantum tasks advancement of quantum algorithms materials science applications
Higham et al. [23]	2023	Improving computational time using QOTA	Quantum computing	Advancement of quantum machine learning quick decision-making or real-time analysis
Yu et al. [24]	2023	Development of a quantum protocol for DNN inference combining quantum oblivious transfer with DNN	Quantum computing and deep neural networks	Enhanced data security for dnns practical deployment of dnns in secure quantum environments
Wen et al. [25]	2024	Integration of auxiliary qubits application of residual connections	Quantum neural networks	Quantum decoherence scalability integration with classical systems
Mellak et al. [26]	2024	CNN model implementation independent parameterization performance improvement	Quantum mechanics	High computational complexity dependency on prior knowledge steady state identification

11.3 Fundamentals of Quantum Machine Learning

Quantum Machine Learning (QML) is new appearing area that integrate the concept of quantum computing with classical machine learning approaches. By leveraging the unique properties of quantum mechanics, such as superposition, entanglement, and quantum parallelism, QML focus to solve complex challenges more accurately than classical methods. Below, we Inquire the main components and applications of QML, including qubits, quantum algorithms, quantum neural networks, quantum reinforcement learning, and quantum clustering techniques.

- Quantum Bits (Qubits): At the core of quantum computing lies the idea of qubits. Unlike classical bits, which are binary and can only be in a state of 0 or 1, qubits can survive in a superposition of two states concurrently. This effects allows quantum systems to process and store exponentially more information than classical systems, activating quickly and more efficient computations for specific tasks.
- Quantum Algorithms: Quantum algorithms are the driving force behind the computational advantages of QML. These algorithms are designed to exploit the parallelism and interference properties of quantum systems, allowing them to solve certain problems, such as database searches or optimization tasks, significantly faster than classical algorithms. This makes them a powerful tool for tackling complex challenges in machine learning.
- Quantum Neural Networks (QNNs): Quantum Neural Networks are a quantum improved version of traditional neural networks. By integrating quantum mechanics principles, QNNs can process and learn from data in ways that classical systems cannot. They leverage quantum superposition and entanglement to explore multiple solutions simultaneously, potentially leading to more accurate predictions and faster learning processes.
- Quantum Reinforcement Learning: Quantum Reinforcement Learning extends classical reinforcement learning by operating in a quantum environment. This approach trains models to make optimal decisions by maximizing rewards derived from quantum interactions. By utilizing quantum parallelism, these models can explore multiple decision paths at once, leading to more efficient and effective learning compared to classical methods.
- Clustering and Pattern Recognition: QML also excels in clustering and pattern recognition tasks, where it can identify hidden structures and relationships in large datasets. Quantum clustering techniques leverage quantum principles to process and analyze data more efficiently than classical unsupervised learning methods, enabling deeper insights and more accurate pattern detection in complex datasets.

11.3.1 Application of Machine Learning Using Quantum Computation

Quantum Machine Learning (QML) is transforming various fields by incorporating quantum computing concepts with classical machine learning techniques. By harnessing the power of quantum mechanics-such as superposition, entanglement, and quantum parallelism-QML enables quickly, more efficient, and innovative solutions to difficult challenges. Below, we explore key applications of QML, ranging from quantum state classification and error correction to optimization, simulation, and beyond, highlighting how these advancements are transforming industries and pushing the boundaries of computational capabilities.

1. Quantum State Classification: Machine learning algorithms can be employed to classify quantum states based on their unique properties. This capability is cru-

cial in quantum information science, where accurately distinguishing between different quantum states is important for applications like quantum communication, cryptography, and quantum error correction. By leveraging ML, researchers can automate and enhance the precision of state identification, enabling more robust quantum systems.

2. Quantum Error Correction: Machine learning techniques can plays a main role in developing error correction codes in quantum computers. Quantum systems are inherently prone to errors due to decoherence and noise. ML can analyze error patterns in quantum computations and help design optimal error correction strategies, thereby improving the reliability and stability of quantum computations. This is a critical step toward building scalable and fault-tolerant quantum computers.
3. Quantum Circuit Optimization: Quantum circuit optimization involves using machine learning to identify the most efficient configurations of quantum gates for specific algorithms. By reducing the number of gates required and minimizing resource usage, ML can enhance the overall performance of quantum algorithms. This optimization is vital for improving the scalability and practicality of quantum computing in real-world applications.
4. Quantum Reinforcement Learning: In quantum reinforcement learning, machine learning is used to train agents that interact with quantum environments. These agents learn to optimize their actions based on rewards derived from their interactions. Quantum reinforcement learning can lead to more efficient learning strategies by leveraging quantum parallelism and superposition, enabling agents to explore multiple decision paths simultaneously and achieve faster convergence.
5. Quantum Data Analysis: Machine learning can be applied to analyze information generated from quantum experiments, which often exhibit complex and non-classical behavior. Assignments such as pattern recognition, anomaly detection, and state characterization in quantum data can be challenging due to the inherent quantum nature of the information. ML techniques provide powerful tools to extract meaningful insights from such data, advancing research in quantum physics and engineering.
6. Hybrid Quantum-Classical Models: Hybrid models incorporate the potential of classical machine learning and quantum algorithms to achieve superior performance in tasks like classification and regression. These models leverage classical ML for preprocessing and postprocessing while utilizing quantum algorithms for computationally intensive tasks. This synergy enables more efficient problem-solving than either classical or quantum systems could achieve independently.
7. Quantum Data Classification: Quantum algorithms excel at classification tasks, where data points must be assigned to predefined categories. The unique structure of quantum systems allows for improved accuracy and efficiency in processing complex datasets. Quantum data classification leverages quantum state-based processing to analyze high dimensional information more effeciently, offering advantages in speed and precision over classical methods.
8. Quantum Feature Extraction: Quantum machine learning enables the extraction of critical features from high-dimensional datasets, which is particularly benefi-

cial in fields like image and signal processing. Traditional methods often struggle with the "curse of dimensionality," but quantum approaches, such as Quantum Support Vector Machines, can process such data more efficiently. These methods exploit quantum states to achieve faster training times and higher accuracy, making them ideal for large-scale data analysis.

9. Optimization Problems: Quantum machine learning provides innovative solutions for difficult optimization problems in finance, logistics, and engineering. Algorithms like the Quantum Approximate Optimization Algorithm (QAOA) and quantum annealers provides suggestive enhancements over classical heuristics. These quantum methods optimize parameters for machine learning models, reducing computation time and enhancing performance in tasks like portfolio optimization and route planning.
10. Enhanced Simulation Models: Quantum machine learning enhances the simulation of quantum systems, enabling more accurate modeling in chemistry, physics, and material science. By leveraging QML, researchers can simulate molecular interactions at the quantum level, accelerating advancements in drug discovery and material design. This capability is critical for understanding complex quantum phenomena and developing new technologies.
11. Quantum Generative Models: Quantum generative models use quantum algorithms to generate new data points that resemble a given training dataset. These models are particularly valuable in applications like drug discovery, where generating novel molecular structures is essential. Quantum generative models leverage the probabilistic nature of quantum mechanics to create diverse and innovative solutions, offering a powerful tool for scientific research and innovation.
12. Quantum Image Recognition: Quantum properties can significantly enhance image recognition systems by improving their efficiency and accuracy. Applications in autonomous vehicles, medical imaging, and security systems benefit from quantum-enhanced image processing, which can handle large datasets and complex patterns more effectively. Quantum image recognition leverages quantum parallelism to achieve faster and more precise analysis, showing the way for enhancements in AI-driven technologies.

11.3.2 Quantum Machine Learning in Military Application

Quantum computing holds transformative potential for military applications, offering unprecedented capabilities in simulation, optimization, cryptography, and decision making. By leveraging the principles of quantum mechanics, such as superposition, entanglement, and quantum parallelism, defense technologies can achieve breakthroughs in material science, secure communications, logistics, and battlefield strategy. Below, we analyze how quantum computing is secure to transforming key areas within the military domain, enhancing operational efficiency, security, and strategic advantage.

1. Quantum Simulations: The ability of quantum computers to replicate complex quantum structures serves vital needs within quantum chemical research. Defense technologies would benefit from the simulated development of new materials for weapon systems that demonstrate distinct properties, such as enhanced durability, stealth capabilities, or energy efficiency. Quantum simulations enable the rapid exploration of material designs, accelerating the development of advanced defense systems.
2. Optimization Problems: Numerous complex optimization problems encountered during military operations exist in domains such as logistics, supply chain management, and mission planning. Quantum approximation optimization algorithms (QAOA), along with other quantum algorithms, solve these challenges more effectively than traditional methods by achieving improved resource allocation, route optimization, and operational planning. This capability ensures more efficient and agile military operations.
3. Cryptography and Secure Communications: Quantum computing systems have the ability to deliver significant improvements to military communication systems. Unrivaled security is achieved through Quantum Key Distribution (QKD), which uses quantum mechanics principles to enable secure key exchange processes. This application ensures that sensitive military data remains protected from interception or decryption by adversaries, safeguarding critical communication channels.
4. Machine Learning and AI Enhancements: Quantum computers amplify machine learning algorithms by speeding up big data processing and analysis. This enhancement meets critical military requirements, such as threat detection, autonomous target examination, and mixed-source information aggregation. By improving situational awareness and decision-making, quantum-enhanced AI systems provide a strategic edge in dynamic and complex operational environments.
5. Battlefield and War Game Simulations: Large-scale battlefield simulation efforts benefit from quantum computing, as it enables the rapid assessment of multiple warfare scenarios. Quantum systems allow military strategists to test different tactical approaches and predict outcomes faster and more accurately than traditional frameworks. This capability supports the development of robust and adaptive military strategies.
6. Signal Processing: Quantum computers boost signal processing efficiency, particularly in electronic warfare applications that require the analysis of broadcast frequencies and signals. Innovations in quantum signal processing lead to improved surveillance data collection, enhanced reconnaissance capabilities, and the development of advanced countermeasure technologies. These advancements strengthen defense systems against adversarial threats.

11.3.3 Quantum Application in Deep Learning Architecture

The integration of deep learning techniques into quantum computing has unlocked up new avenues for optimizing quantum circuits, improving data representations, and enhancing computational efficiency. By leveraging the power of neural networks and reinforcement learning, researchers can explore novel ways to design, analyze, and optimize quantum systems. Below are some key applications of deep learning in quantum computing:

1. Quantum Circuit Learning: Deep learning methods assist researchers in designing optimal quantum circuits tailored to specific computational needs. By training neural networks to explore different quantum circuit configurations, deep learning helps identify the most effective setups for desired quantum operations.
2. Enhancing Quantum Representations: Deep learning models play a important role in encoding and decoding complex quantum information, leading to optimal representations of quantum states. This advancement significantly benefits quantum communication and cryptography by improving security and efficiency in information transfer.
3. Quantum Classifiers: Deep learning enables the training of quantum models to function as classification systems capable of analyzing quantum data. Methods such as quantum-enhanced support vector machines (SVMs) leverage deep learning to improve precision and scalability in quantum classification tasks.
4. Quantum Image and Signal Processing: Deep learning enhances image and signal processing in quantum systems, providing the way for quantum improved imaging technology and advanced multi-qubit signal analysis. These improvements have real-world applications in medical imaging, remote sensing, and secure quantum communications.
5. Deep Reinforcement Learning for Quantum Control: Quantum system control can be significantly improved through deep reinforcement learning, which optimizes strategies based on system feedback and dynamic gate operations. This approach enhances the precision and adaptability of quantum computing processes.
6. Quantum Graph Neural Networks: Extending graph neural networks to quantum systems enables the analysis of complex quantum interactions and network structures. These models provide insights into the relationships between quantum states, improving predictions and optimizations in quantum computing applications.
7. Modeling Quantum Dynamics: Deep learning facilitates efficient simulations of quantum dynamic processes by modeling the time evolution of quantum systems. This capability allows researchers to explore quantum behaviors that were previously computationally infeasible.
8. Causal Inference in Quantum Mechanics: Deep learning aids in uncovering causal relationships in quantum experiments, shedding light on foundational aspects of quantum mechanics. By identifying dependencies between quantum events, researchers can refine their understanding of quantum causality.

9. Optimizing Quantum Sampling: Quantum sampling methods, essential for quantum Monte Carlo simulations, can be optimized using deep learning models. These enhancements improve the efficiency and accuracy of quantum sampling techniques, benefiting computational physics and quantum chemistry applications.
10. Quantum Feature Selection: Deep learning helps identify the most relevant quantum data features for machine learning tasks, ensuring more effective utilization of quantum information. This leads to better performance in quantum machine learning applications and hybrid quantum-classical models.

11.4 Neural Networks in Quantum Computation

Quantum neural networks (QNNs) have surfaced as a promising route to revolutionizing artificial intelligence and machine learning by exploiting the methods of quantum mechanics. One of their primary advantages is computational efficiency, particularly when dealing with more-dimensional data and complex patterns. Unlike classical neural networks, QNNs can exploit quantum superposition and entanglement to process information in parallel, enabling them to perform certain computations significantly faster. These capabilities position QNNs as potential game-changers in tasks that demand substantial computational power, such as optimization, pattern recognition, and big data analysis. The structural architecture for quantum neural network is shown in Fig. 11.1.

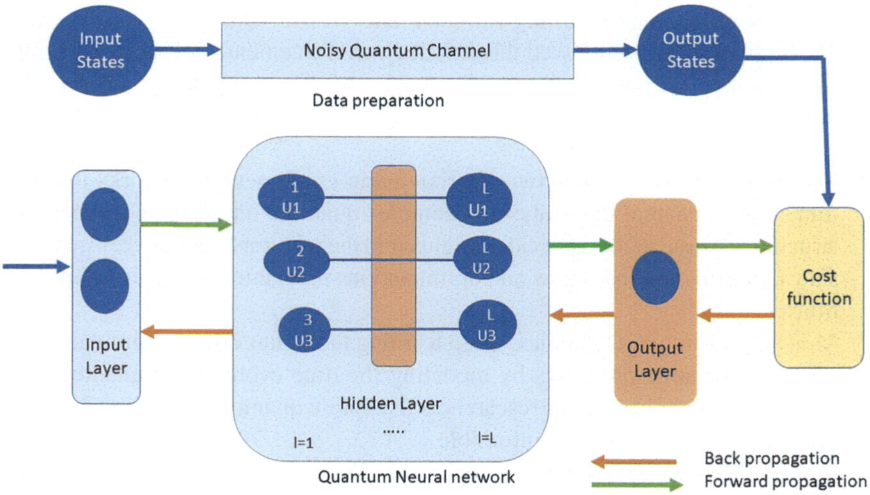

Fig. 11.1 Quantum neural network

In addition to their speed, QNNs demonstrate strong potential in handling complex problems, particularly in reinforcement learning scenarios. Many real-world problems, such as those modeled by Markov decision processes, require efficient decision-making frameworks. QNNs can enhance reinforcement learning algorithms by processing complex states more effectively and potentially finding optimal solutions with fewer iterations. However, like classical deep learning models, QNNs are also susceptible to fundamental challenges such as the vanishing gradient problem. Addressing this issue in quantum environments requires innovative approaches, as traditional techniques used in classical deep learning may not be directly applicable. Another significant challenge in developing QNNs is their compatibility with current quantum hardware, particularly noisy intermediate scale quantum (NISQ) devices. The limitations of these devices, such as restricted qubit counts and high error rates, hinder the practical implementation of QNNs. Researchers are actively working on optimizing quantum algorithms that can function effectively within the constraints of NISQ hardware, facilitate for practical usage of applications of QNNs in the near term. Despite these challenges, the potential applications of QNNs span various fields, including communication networks, the Internet of Things (IoT), and blockchain technology. By integrating QNNs, these systems could experience significant advancements in computational efficiency, security, and scalability.

QNNs also hold promise in specialized domains such as quantum machine learning, where they can outperform classical neural networks in classification, regression, and clustering tasks. They are expected to improve reinforcement learning frameworks, making them valuable for optimizing decision-making in complex environments. Furthermore, in quantum chemistry and material science, QNNs can efficiently model quantum systems, aiding in molecular property prediction, chemical reaction studies, and new material discoveries. Their applicability extends to combinatorial optimization problems in logistics, finance, and operations research, where quantum algorithms can find optimal solutions more rapidly than classical methods. In communication networks, QNNs can enhance data processing and resource allocation, benefiting distributed systems such as blockchain and IoT networks. Lastly, QNNs can revolutionize image and signal processing by replacing classical convolutional neural networks (CNNs) in tasks requiring pattern recognition, showing superior performance in computer vision and audio processing applications. As research in quantum computing advances, the incorporation of QNNs into real world applications will likely drive major breakthroughs across multiple industries.

Quantum state embedding is a foundational component of Quantum Neural Networks (QNNs), where quantum data is seamlessly integrated into neural network architectures to amplify their computational power. By encoding information directly into quantum states, QNNs can execute sophisticated transformations and uncover intricate patterns that classical networks cannot access. This approach advantage the uncommon properties of quantum mechanics, such as superposition and entanglement, to enable highly efficient quantum-enhanced models. These models are particularly valuable in applications like quantum cryptography, secure communications, and machine learning tasks that rely on deep quantum correlations. As advancements in quantum embedding techniques continue to evolve, the fusion of quantum

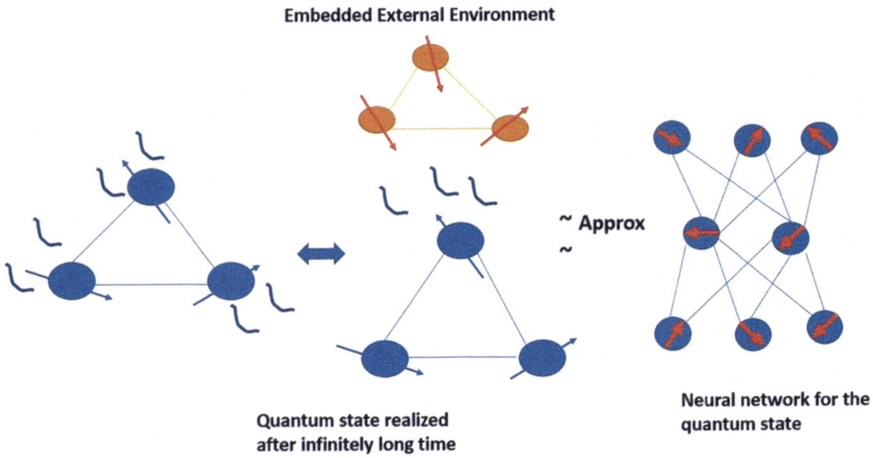

Fig. 11.2 Neural network with quantum states

mechanics and artificial intelligence is poised to unlock groundbreaking possibilities in computational science, paving the way for transformative innovations across various domains. The improvised neural network embedding with quantum states is shown in Fig. 11.2.

11.5 Challenges and Opportunities

Quantum Neural Networks (QNNs) present both unique challenges and exciting opportunities as they evolve within the broader field of quantum machine learning (QML). While they hold great promise for computational advancements, significant hurdles must be addressed before they can be effectively implemented at scale.

11.5.1 Challenges

- Scalability: As quantum systems grow in size, the complexity of quantum algorithms also increases. A major obstacle in developing QML algorithms is ensuring they can scale effectively to larger datasets and more intricate models without becoming computationally infeasible.
- Limited Quantum Resources: Current quantum computers have a restricted number of qubits and available quantum gates, restricting the complexity of quantum circuits that can be executed. This limitation affects the richness and expressiveness of QNN models, making it difficult to develop deep and complex quantum networks.

- Training Complexity: Training QNNs is inherently more challenging than training classical neural networks due to the nature of quantum states and measurements. The process often requires frequent evaluations, which can be resource-intensive and slow down convergence.
- Lack of Theoretical Framework: The theoretical foundations for QML and QNNs are still underdeveloped. There is a pressing need for a solid theoretical framework that can guide the creation of efficient quantum algorithms and establish clear benchmarks for their performance.
- Data Encoding: Converting classical data into quantum states is a suggestive challenge. The choice of encoding methods has a major impact on the performance of QML algorithms, and identifying optimal encoding techniques remains an active area of research.
- Overfitting: Similar to traditional machine learning models, QNNs are susceptible to overfitting, especially when trained on limited datasets. Developing techniques to prevent overfitting in quantum environments is crucial for ensuring robust generalization to unseen data.

11.5.2 Opportunities

- Enhanced Computational Power: QML algorithms have the power to solve challenges that classical computers struggle with. By advantage quantum superposition and entanglement, QNNs can process vast amounts of data in parallel, leading to faster and more efficient computations.
- Improved Data Analysis: Quantum algorithms can significantly improve data analysis, particularly in high-dimensional spaces. This has important implications for fields such as finance, healthcare, and scientific research, where handling large and complex datasets is critical.
- Quantum Feature Learning: QNNs excel in learning and extracting features from quantum data, improving machine learning performance in domains like quantum mechanics, quantum chemistry, and material science. These capabilities are crucial for simulating and understanding complex quantum phenomena.
- Hybrid Quantum Classical Models:Incorporating quantum and traditional models can maximize the strengths of both paradigms. Hybrid approaches offer enhanced performance for applications such as optimization, simulations, and machine learning tasks that require both quantum speedup and classical robustness.
- Optimization Problems: Many real world challenges, such as logistics, supply chain management, and resource allocation, can be formulated as optimization problems. QML techniques offer the power to solve these problems more efficiently using quantum optimization algorithms.
- Pattern Recognition and Classification: QNNs are wellsuited for pattern recognition and classification tasks, especially in highdimensional domains. Their appli-

cations range from image and voice recognition to anomaly detection in various industries.

- Real Time Decision Making: The speed of quantum computations could enable real time decision making in dynamic and rapidly changing environments, such as stock trading and artificial intelligence applications. QML could provide instant insights, improving decision-making processes across multiple domains.

11.6 Future Research Directions

The field of Quantum Neural Networks (QNNs) presents numerous opportunities for future research, focusing on improving training methodologies, hybrid quantum classical models, data storage, and specialized applications. These research directions aim to interface the rift between theoretical enhancements and practical execution, paving the way for enhanced machine learning and artificial intelligence capabilities.

1. Quantum Aid for Training and Evaluation

    - Gradient Computation: Training quantum neural networks involves complex computations, often requiring backpropagation or gradient descent. Future research could focus on leveraging quantum feedback algorithms to enhance these processes, making matrix multiplication and optimization more efficient.
    - Quantum Search Optimization: Grover-like quantum search algorithms can be explored to optimize hyperparameters and select network weights more efficiently than classical methods. This could significantly improve training efficiency and model performance.
    - Quantum Feature Encoding: As datasets continue to grow in complexity, quantum states can be utilized to encode high-dimensional data more efficiently. Quantum encoding methods could provide exponential storage advantages and enable faster processing of large feature spaces.

2. Hybrid Models of Quantum and Classical Networks

    - Intermediate Quantum Layers: Integrating quantum layers within classical neural networks can enhance specific tasks, such as feature extraction and kernel computation. These hybrid architectures could lead to more potent and flexible machine learning models.
    - Data Preprocessing: Quantum circuits can be employed in data preprocessing pipelines before classical neural networks process the information. This could improve data transformation and representation, enabling better performance in downstream tasks.

3. Quantum Memory and Data Storage

- Efficient Storage: Quantum memory offers a potential breakthrough in storing vast amounts of data in compact quantum states. This could revolutionize big data applications by enabling efficient compression and retrieval mechanisms.
- Quantum Access Patterns: Quantum Random Access Memory (QRAM) could enable highly efficient searching and retrieval of large datasets, enhancing computational speed in data-intensive tasks.
- Entangled Data Representations: Utilizing quantum entanglement to represent relationships between data points could introduce novel approaches in data augmentation and relational learning, leading to improved machine learning performance.

4. Quantum Advantage in Specific Applications

- Quantum Natural Language Processing (QNLP): The development of QNNs tailored for natural language processing assignment, such as modeling sentence structures and semantic embeddings, could lead to more advanced AI-driven language understanding.
- Quantum Reinforcement Learning: By leveraging quantum principles, reinforcement learning models can improve decision-making in probabilistic and uncertain environments, offering potential applications in robotics, finance, and automated systems.
- Generative Models: Quantum-enhanced generative models could improve sample generation quality, leading to advancements in fields such as image synthesis, data augmentation, and simulation-driven applications.

11.7 Conclusion

Quantum neural network is an innovative concept the integration of quantum computing and deep learning suggests new broad technology breakthroughs in the field of data processing. Much work remains in their development even so QNNs could enable features like exponential accelerations, superior representation capability, and increased model applicability. The application of quantum principles in neural networks can change industries from natural language processing to generative modeling, and reinforcement learning. However, for this vision to become reality, several concrete problems have to be solved; noise in the quantum hardware, scalability, and creating efficient hybrid quantum classical systems. It is predicted that as research goes on the interconnection between quantum computing and artificial intelligence will revolutionize the fundamental advanced computation in a way that will not only customize the AI breakthroughs but also increase the understanding of theoretical principles of both quantum computing and AI. QNNs are destined to play an important role of the AI systems belonging to the next generation numerous opportunities are opening for experimenting.

References

1. Yousif, M., Al-Khateeb, B., Garcia-Zapirain, B.: A new quantum circuits of quantum convolutional neural network for x-ray images classification. IEEE Access (2024)
2. Priyanka, G., Venkatesan, M., Prabhavathy, P.: Advancements in quantum machine learning and quantum deep learning: a comprehensive review of algorithms, challenges, and future directions, pp. 1–8. IEEE (2023)
3. Liang, Y., Peng, W., Zheng, Z.-J., Silvén, O., Zhao, G.: A hybrid quantum-classical neural network with deep residual learning. Neural Netw. **143**, 133–147 (2021)
4. Le, L., Nguyen, T.N.: Dqra: Deep quantum routing agent for entanglement routing in quantum networks. IEEE Trans. Quantum Eng. **3**, 1–12 (2022)
5. Trochun, Y., Pavlov, E., Stirenko, S., Gordienko, Y.: Impact of hybrid neural network structure on performance of multiclass classification, pp. 152–156. IEEE (2021)
6. Lokes, S., Mahenthar, C.S.J., Kumaran, S.P., Sathyaprakash, P., Jayakumar, V.: Implementation of quantum deep reinforcement learning using variational quantum circuits, pp. 1–4. IEEE (2022)
7. Han, X., et al.: Combining graph neural network with deep reinforcement learning for resource allocation in computing force networks. Front. Inf. Technol. Electron. Eng. **25**, 701–712 (2024)
8. Akash, A.R., et al.: Quantum convolutional neural network-based online malware file detection for smart grid devices, pp. 1–5. IEEE (2023)
9. Gupta, S., Mohanta, S., Chakraborty, M., Ghosh, S.: Quantum machine learning-using quantum computation in artificial intelligence and deep neural networks: quantum computation and machine learning in artificial intelligence, pp. 268–274. IEEE (2017)
10. Kawase, Y.: Distributed quantum neural networks via partitioned features encoding. Quantum Mach. Intell. **6**, 15 (2024)
11. Hdaib, M., Rajasegarar, S., Pan, L.: Quantum deep learning-based anomaly detection for enhanced network security. Quantum Mach. Intell. **6**, 26 (2024)
12. Ratnaparkhi, A.A., Pilli, E., Joshi, R.: Survey of scaling platforms for deep neural networks, pp. 1–6. IEEE (2016)
13. Golchha, R., Verma, G.K.: A deep learning model for multiclass image classification using quantum cnn, pp. 102–107. IEEE (2023)
14. Incudini, M., et al.: The quantum path kernel: a generalized neural tangent kernel for deep quantum machine learning. IEEE Trans. Quantum Eng. **4**, 1–16 (2023)
15. Chen, S.Y.-C.: tQuantum deep q-learning with distributed prioritized experience replay, vol. 2, pp. 31–35. IEEE (2023)
16. Kwak, Y., Yun, W.J., Jung, S., Kim, J.: Quantum neural networks: concepts, applications, and challenges, pp. 413–416. IEEE (2021)
17. Manjunath, T., Bhowmik, B.: Quantum-enhanced deep q learning with parametrized quantum circuit, pp. 1–6. IEEE (2024)
18. Padha, A., Sahoo, A.: Quantum deep neural networks for time series analysis. Quantum Inf. Process. **23**, 205 (2024)
19. Bova, F., Goldfarb, A., Melko, R.G.: Commercial applications of quantum computing. EPJ Quantum Technol. **8**, 2 (2021)
20. Krelina, M.: Quantum technology for military applications. EPJ Quantum Technol. **8**, 24 (2021)
21. Moussa, C., Wang, H., Bäck, T., Dunjko, V.: Unsupervised strategies for identifying optimal parameters in quantum approximate optimization algorithm. EPJ Quantum Technol. **9**, 11 (2022)
22. Li, J., Kais, S.: Quantum cluster algorithm for data classification. Mater. Theory **5**, 1–14 (2021)
23. Higham, C.F., Bedford, A.: Quantum deep learning by sampling neural nets with a quantum annealer. Sci. Rep. **13**, 3939 (2023)
24. Yu, H., Ren, X., Zhao, C., Yang, S., McCann, J.: Quantum-aided secure deep neural network inference on real quantum computers. Sci. Rep. **13**, 19130 (2023)
25. Wen, J., Huang, Z., Cai, D., Qian, L.: Enhancing the expressivity of quantum neural networks with residual connections. Commun. Phys. **7**, 220 (2024)

26. Mellak, J., Arrigoni, E., von der Linden, W.: Deep neural networks as variational solutions for correlated open quantum systems. Commun. Phys. **7**, 268 (2024)
27. Chen, H.-Y., Chang, Y.-J., Liao, S.-W., Chang, C.-R.: Deep q-learning with hybrid quantum neural network on solving maze problems. Quantum Mach. Intell. **6**, 2 (2024)

The manufacturer's authorised representative in the EU is Springer Nature Customer Service Centre GmbH, Europaplatz 3, 69115 Heidelberg, Germany. If you have any concerns regarding our products, please contact ProductSafety@springernature.com

Printed and bound by CPI Group (UK) Ltd, Croydon, CR0 4YY
01/10/2026
02225850-0006